High-Level Radioactive Waste Management

High-Level Radioactive Waste Management

Milton H. Campbell, EDITOR
Exxon Nuclear Company, Inc.

A symposium sponsored by the Division of Nuclear Chemistry and Technology at the 167th Meeting of the American Chemical Society, Los Angeles, Calif., April 1–2, 1974.

ADVANCES IN CHEMISTRY SERIES **153**

AMERICAN CHEMICAL SOCIETY
WASHINGTON, D. C. 1976

Library of Congress CIP Data

High-level radioactive waste management.
(Advances in chemistry series; 153. ISSN 0065-2393)

Includes bibliographical references and index.

1. Radioactive waste disposal—Congresses.
I. Campbell, Milton H., 1928- . II. American Chemical Society. Division of Nuclear Chemistry and Technology. III. Series: Advances in chemistry series; 153.

QD1.A355 no. 153 (TD898) 540'.8s
(621.48'38) 76-25020
ISBN 0-8412-0270-2 ADCSAJ 153 1-166 (1976)

PRINTED IN THE UNITED STATES OF AMERICA

Advances in Chemistry Series

Robert F. Gould, *Editor*

FOREWORD

ADVANCES IN CHEMISTRY SERIES was founded in 1949 by the American Chemical Society as an outlet for symposia and collections of data in special areas of topical interest that could not be accommodated in the Society's journals. It provides a medium for symposia that would otherwise be fragmented, their papers distributed among several journals or not published at all. Papers are refereed critically according to ACS editorial standards and receive the careful attention and processing characteristic of ACS publications. Papers published in ADVANCES IN CHEMISTRY SERIES are original contributions not published elsewhere in whole or major part and include reports of research as well as reviews since symposia may embrace both types of presentation.

CONTENTS

PREFACE

High-level radioactive waste management has become a topic of discussion throughout the scientific and engineering communities, a subject for our nation's newspapers to review for the lay public, and a major issue for interveners in the nuclear program. While many papers of good technical quality, but of limited scope, have been written on this subject, there is presently available no single source that presents an overview of high-level radioactive waste management. This volume of the *Advances in Chemistry Series* is intended to fill the gap.

High-level radioactive waste management covers a broad area of federal government and commercial activities. Chronologically, the first wastes of this nature were generated in the Manhattan Project of World War II vintage. The U.S. Atomic Energy Commission (now the Energy Research and Development Administration), the outgrowth of the Manhattan Project, pursued two simultaneous courses after World War II. On the one hand, support for our national defense plans required plutonium, while on the other hand, a program for peaceful use of atomic energy promoted an increasingly sophisticated nuclear technology devoted to production of electrical power. A common product of both courses is high-level radioactive waste. However, the waste takes a multitude of forms and has no conveniently simple definition.

There are several natural subdivisions of the overall topic that have been considered in organizing this volume. First, the Atomic Energy Commission's philosophy and consequent policy are presented in Dr. Frank K. Pittman's paper titled "Management of Commercial Radioactive Waste."

These papers describe the programs directed to high-level radioactive waste management at the Atomic Energy Commission sites where the waste was generated in support of AEC programs. In this context, "Solid Waste Forms for Savannah River Plant Radioactive Wastes" are discussed in some depth by R. M. Wallace et al. C. M. Slansky describes the "High-Level Waste Management Program" at the National Reactor Testing Station. W. W. Schulz and M. J. Kupfer review "Solidification and Storage of Hanford's High-Level Radioactive Liquid Wastes." These three papers present in considerable detail the aggressive programs being actively pursued at the Atomic Energy Commission sites.

Two papers about high-level waste management plans for commercial reprocessing plants complete the overview of operating plant activities. J. P. Duckworth details the Nuclear Fuel Services plans. R. G.

Barnes et al. describe the Midwest Reprocessing Plant projections in depth.

There is also a significant research effort directed toward the long-term solution of high-level radioactive waste management. J. O. Blomeke and W. D. Bond give an overview of the Oak Ridge National Laboratory program, while J. E. Mendel et al. present the scope of Battelle Pacific Northwest Laboratories activities. Further papers present topical discussions of significant individual investigations. G. S. Barney describes a unique hydrothermal reaction for incorporating caustic wastes in a synthetic mineral; S. Fried et al. tell of unusual techniques they have developed to assess the distribution of plutonium in rock containment environments; M. W. Wilding and R. W. Rhodes report on a process for removing cesium and strontium from fuel storage basin water. N. W. Golchert addresses the important consideration for measuring transuranics in environmental water. The overview is completed by the description of waste management practices in Europe presented by R. M. Walton, Jr.

This volume was not assembled simply to reassure the reader, or to project the thought that all the best solutions have been identified and are being implemented. Rather, the intent is to report responsibly to our professional peers the status of this technology on the occasion of the symposium in 1974. Your editor and most of the authors are still active in the field. They stand fully ready to provide information on recent work and continue a constructive dialogue with any reader. Further, we solicit responsible recommendations for alternative solutions; and your editor pledges a professional consideration for such proposals.

Exxon Nuclear Co., Inc.
Richland, Wash.
December 1974

MILTON H. CAMPBELL

1

Management of Commercial High-Level Radioactive Waste

FRANK K. PITTMAN

Division of Waste Management and Transportation,
U. S. Atomic Energy Commission, Washington, D. C. 20545

There is now available the technology which will allow us to contain the commercial high-level radioactive waste from the nuclear power industry. Through proper waste management controls we can reduce to acceptable levels the potential for public hazard and environmental damage. The AEC's program will provide retrievable storage for these solidified wastes in carefully maintained and monitored engineered facilities for the next several decades. During this period investigation and demonstration of waste disposal in a deep, stable, geologic formation will continue. After the completion of an extensive testing and demonstration program establishing the acceptability of the method of permanent disposal, the radioactive waste will be removed from the retrievable surface storage facility and disposed of permanently.

The disposition of radioactive waste from the nuclear electric power industry continues to be a subject of much interest to the public, and in particular, to those members of the public searching for any possible vehicle to stop or delay the progress of the nuclear power program. A vast reservoir of misunderstanding and misinterpretation prevails in the mind of the lay public. This is often fanned into almost hysterical fears by a few technical people who should recognize that the protection of the public from the potential dangers of the radioactive waste is by no means an insurmountable problem. The major factor in this highly emotional issue is the failure of the public, and some of its more vocal "scientific" advisors, to recognize that there is available now the technology which will allow us to contain the waste in a way that will reduce the potential for environmental damage and public hazard to acceptable levels.

We are continuously faced with the assertion that presently we do not know how to manage permanently the waste, and that, until we do, activities to generate it should not be permitted. Analysis of such assertions shows that what is being said is that since we are not today able to "neutralize" the radioactivity or remove it completely from the total earth environment, we do not know what to do with it.

This is fallacious reasoning. The mere existence of radioactive material in the total earth environment is not the important factor. Rather, the question is: can it be isolated from the biosphere upon which man depends for his existence. The answer to this question is yes, and it makes no difference, insofar as the health and safety of man is concerned, whether this isolation is brought about by continuously monitored and maintained containment on the surface of the earth or by unmaintained containment in massive geologic formations beneath the earth's surface. There may be economic reasons why one method is preferable, but there are no safety reasons.

The AEC (now the Energy Research and Development Administration, ERDA) program over the past years, has shown that either method can be used, as far as technology is concerned. Initial monitored surface storage has the distinct advantage of keeping the radioactive material in an easily retrievable condition. \Should some of the promising technologies for nuclear "neutralization" of the long-lived radioisotopes (transmutation) or for complete removal from the total earth environment (extraterrestrial disposal) develop in the next few decades, the retrievably stored waste can be removed from its storage location and disposed of by these methods. Should these technologies not live up to their promise, or should there arise any reason why surface storage should be discontinued prior to final answers on the long-range methods, the material in retrievable surface storage can always be quickly transferred to selected geologic formations for subsequent isolation.

There is no reason why decisions should be made today concerning the method of control and containment that might be used a hundred years from now since we know today how to safely contain and isolate the radioactive material for such time periods in a way that will not in any way preclude subsequent management of the waste either by geologic storage, which we are now sure can be used, or by other techniques which are potentially more attractive but about which there are still many technical questions.

The AEC will, as announced in May of 1972, manage the commercial solidfied high-level waste by storage in surface facilities at an AEC-owned site. The stored containers of waste will be monitored continuously, and they can be easily removed from storage for maintenance and repair if the monitoring activity indicates that this is necessary.

The structures used for receiving, cleaning, repackaging, or overpacking, as required, and for repair and maintenance of waste canisters (as well as those connected with the actual storage itself) will be designed to facilitate their maintenance and repair. They can be replaced, if necessary, without in any way introducing safety problems to the storage activities.

For the past two years, we have been evaluating the various engineering techniques that can be used to store radioactive waste in surface facilities for extended periods of time. I will review briefly the status of this evaluation.

Before doing this, however, I would like to emphasize that extended storage of radioactive waste material introduces some problems that do not exist for short-term storage required for spent fuel awaitiing processing or even solidified waste in storage at the spent fuel processors' sites for up to 10 years awaiting transfer to the AEC. Techniques which might be perfectly acceptable for such interim storage would not necessarily be optimum or even acceptable for extended storage.

With this in mind, let's look at what is required for safe extended storage of high-level radioactive waste, and how the various concepts we have considered will meet these requirements.

Very simply stated the only things we must accomplish in storage are: (1) prevention of the radioactive waste from escaping to the environment; and (2) protection of the operating personnel and the public from penetrating radiation while the radioactive material is being stored. Storage is a passive, not an active operation.

Protection against penetrating radiation, regardless of the storage concept or its location, is accomplished by isolation, shielding, or a combination of the two. Standard shielding materials—concrete, steel, water, earth, uranium, lead, etc.—can be used, and the choice is more or less one of economics.

The basic techniques for preventing the waste from escaping from the container are also quite simple. All that must be accomplished is to prevent containment failure caused by: (1) internal pressure buildup; (2) various chemical, grain boundary, stress, galvanic and other forms of cororsion, both internal and external; (3) weld failure; (4) excessive heat; (5) radiation damage; or (6) physical forces. The basic challenge of safe surface storage is to assure, first, that containment failures are kept to an absolute minimum, and second, that when minor failures do occur, the system of surveillance will detect the failure in time to allow the failed container to be retrieved and repaired before unacceptable amounts of radioactive material have escaped to the biosphere.

When we examine the various forces which can be instrumental in containment failure, we find that certain of them result from the form of the contained waste, others from the geometry, construction materials, and quality of fabrication and welding of the container itself, others from the specific approach to storage that is used, and still others from the characteristics of the site at which storage takes place. Naturally, we must, and will, take full advantage of all technology available now; as future technology is developed, we will see that it is applied.

In this paper, I will limit my discussion to those forces which result from the engineering concept used for storage and from the location of the storage activity. In general, the rupture forces which depend on waste form and container characteristics do not influence the selection of storage concept or site.

When we initiated the program for evaluating the various engineering options for surface storage, it was evident that there were several factors to be considered, each having an impact on the safety, efficiency, and acceptability of a storage concept. In general, these are:

1. Method of heat removal
2. Method of shielding
3. Interaction of coolant and waste containers
4. Reliability of mechanical equipment
5. Method(s) of surveillance of waste containers
6. Methods of removal of waste containers
7. Methods of maintenance of waste containers and structures
8. Construction and operating costs
9. Vulnerability to natural or man-made catastrophic events

The major factor is the coolant to be used to remove the heat generated by the radioactive decay of the waste. Each canister of waste will generate 2–5 kW of heat when it is delivered for storage. This heat will, of course, decrease with time, essentially on a half-life of 30 years since the principal heat generators are ^{90}Sr and ^{137}Cs with half-lives in that range. For the purposes of design, planning, and evaluation, we have assumed that the initial heat load of each reference canister ($1'D \times 10'L$) will be 5 kW.

Early in the project, we decided to limit the design evaluation to concepts using either mechanically circulated water, or mechanically or convectively circulated air. We also examined the question of storage of multiple canisters in modular cells or basins vs. isolated storage of single canisters. Finally, we evaluated storage of the waste in the "as received" condition vs. recanning or overpackaging.

After preliminary evaluation of many combinations of these and other variables, we decided that we would make a conceptual design study of a modular, forced circulation, water-cooled system in which

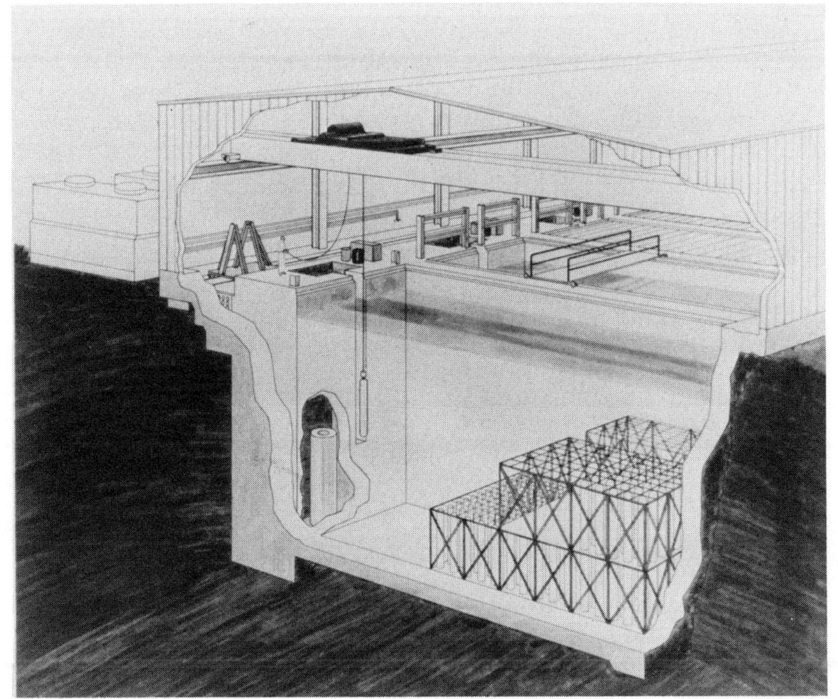

Figure 1. Retrievable surface storage facility, water basin concept, cutaway view

500 canisters—both "as received" and overpacked—would be stored in steel-lined, water-filled, concrete modules. Each module would have its own pump, heat exchanger, and cooling tower. This concept, shown in Figure 1, was established as the "reference design" against which other concepts could be evaluated.

Following that, we developed another alternative conceptual design for a modular, natural circulation, air-cooled vault concept in which overpacked canisters would be stored in steel-lined, concrete vaults through which air is convectively circulated. Figure 2 shows this approach.

We have also developed conceptual designs for several variations of an approach in which one to three "as received" canisters is scaled in mild steel casks with wall thickness between 2 in. and 16 in. surrounded by concrete sleeves of varying thickness to shield against gamma and neutron radiation. The casks with the concrete shielding would be stored in open air and cooling would be accomplished by convective circulation of air between the concrete sleeve and the sealed steel cask as shown in Figure 3. We are now in the final stages of evaluating the several concepts, and a decision will be made in the near future on which approach will be used.

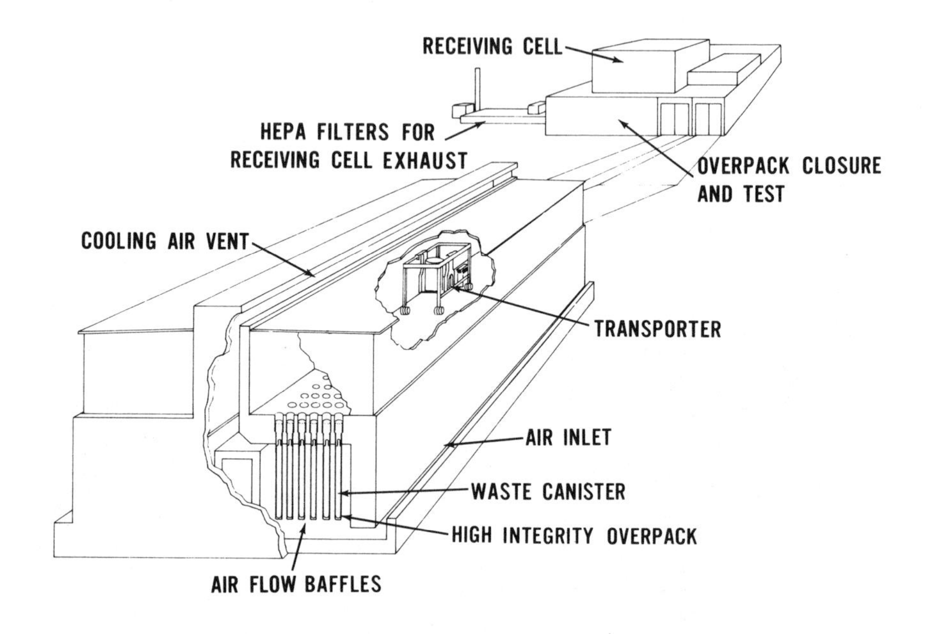

Figure 2. RSSF air cooled vault concept

I would like now to discuss the procedures and criteria used in evaluating the many sites which were originally considered potentially acceptable for surface storage of high-level radioactive waste. Many factors were considered in the preliminary site evaluation. Among the most important were: (1) relative construction and operating cost; (2) AEC or other government ownership; (3) distance from present and expected future locations of spent fuel processing plants; (4) geology, hydrology, seismology, climate, and soil characteristics; (5) availability of multiple modes of transport; (6) isolation; (7) available acreage; (8) availability of power and water; and (9) availability of an adequate manpower force. In addition, and of great importance are: the availability of facilities in which developmental, trouble-shooting, and process and equipment improvement activities could be carried out; as well as the scientific, engineering, and technical talents needed to man these facilities. Another important factor is the likelihood that other interesting and challenging nuclear activities would continue at the location to assure long-term availability of technical and engineering support for the storage activity which in itself probably could not attract the high caliber of scientific and engineering talent needed to assure continued improvement of storage techniques. And, of course, a most important factor in final evaluation is acceptance by the local populace as well as by local and regional political leaders.

Site evaluation efforts have narrowed the potentially attractive locations to three or four of the major AEC sites. We are now in the last stages of evaluation among these sites, and a choice will be made at the same time as selection of the concept to be used for storage.

Earlier in this discussion, I pointed out that an important factor in assuring that containment rupture and its consequences will be minimized is the form of the waste in the container during extended storage. Solidification of aqueous high-level waste by evaporation–calcination techniques gives a waste form which meets the Commission regulation as expressed in 10 CFR 50 Appendix F. This regulation, however, was developed on the assumption that within five to ten years after solidification, the waste would be sealed away in a bedded salt formation where migration of the calcine waste is so slow that no radioactive material could reach the biosphere. The suitability of this waste form for extended surface storage when continued container integrity gives the sole protection against migration to the biosphere is questionable. We are, therefore, now developing techniques of converting calcine, or directly con-

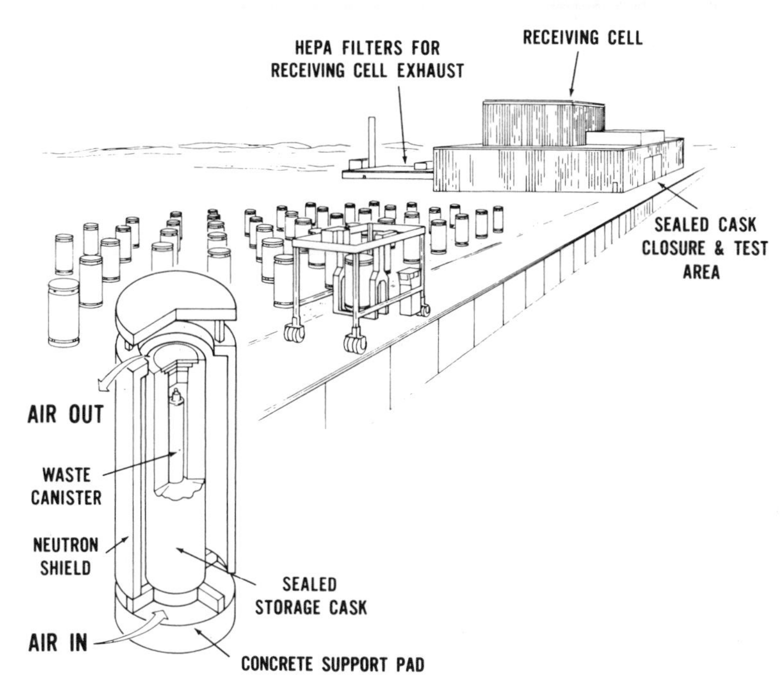

Figure 3. RSSF sealed storage cask concept

verting aqueous radioactive wastes, to a massive, low-leachable form such as glass. We are also analyzing several approaches which could be used to assure that commercial high-level waste in extended storage will be a glass or other equally acceptable form. While I am not, at present, able to discuss these alternative approaches, I do expect that a positive resolution to the problem will be developed in the next few months.

In conclusion, I would like to discuss our recently reoriented and expanded program to develop acceptable techniques for permanent disposal, as opposed to extended surface storage, of the commercial high-level radioactive waste. As we have progressed on our program for surface storage, we have become completely convinced that this approach, which we will use to manage the high-level waste delivered from the initial spent fuel processing effort, can be used for extended periods of time to safely manage the waste. As mentioned earlier, there is no real urgency for us to develop a permanent disposal system. We have, therefore, decided that rather than place all our effort, as we have for the past few years, on the use of bedded salt for ultimate disposal, we would expand our study, evaluation, and experimental effort to other geologic formations which have characteristics that make them potentially as good as, or possibly even better than, bedded salt for this purpose. We are now analyzing such formations as granites, limestones, dolomites, shales, gneiss, schist, and mudstones, as well as domed salt and salt anticlines, to establish our program of investigation. Using the techniques which we developed in connection with our bedded salt work, we will bring our knowledge on one or more of these formations to the same point as that on bedded salt. This will allow us in a few years, if we so desire, to proceed with the in situ pilot emplacement of waste in the formation which has been shown (by a comparative analysis of experimental data rather than by well founded, but subjective, judgment) to be the best for disposal of radioactive waste.

RECEIVED November 27, 1974.

2

Solid Forms for Savannah River Plant Radioactive Wastes

R. M. WALLACE, W. H. HALE, R. F. BRADLEY, H. L. HULL, J. A. KELLEY, J. A. STONE, and G. H. THOMPSON

Savannah River Laboratory, E. I. du Pont de Nemours & Co., Inc., Aiken, S.C. 29801

Methods are being developed to immobilize Savannah River Plant wastes in solid forms such as cement, asphalt, or glass. ^{137}Cs and ^{90}Sr are the major biological hazards and heat producers in the alkaline wastes produced at SRP. In the conceptual process being studied, ^{137}Cs removed from alkaline supernates, together with insoluble sludges that contain ^{90}Sr, will be incorporated into solid forms of high integrity and low volume suitable for storage in a retrievable surface storage facility for about 100 years, and for eventual shipment to an off-site repository. Mineralization of ^{137}Cs, or its fixation on zeolite prior to incorporation into solid forms, is also being studied. Economic analyses to reduce costs and fault-tree analyses to minimize risks are being conducted. Methods are being studied for removal of sludge from (and final decontamination of) waste tanks.

The Savannah River Plant, operated by the Du Pont Co. for the Atomic Energy Commission (now the Energy Research and Development Administration, ERDA) occupies an area of ~300 sq mi along the Savannah River near Aiken, S.C., approximately 22 mi downstream from Augusta, Ga. The plant includes a nuclear reactor fuel fabrication plant, three operational reactors (formerly five), two fuel reprocessing plants, and a facility for producing heavy water. The plant started producing nuclear materials for the AEC in 1953 and has been in continuous operation since.

The radioactive wastes from plant operations are stored as liquids or slurries in double-walled carbon steel tanks. Each tank has a capacity of 750,000–1,300,000 gal. Until recently, the only alternative under serious consideration was storage as a liquid in a deep mined cavern under the plantsite. In November 1972, the AEC deferred further study of that project in favor of an investigation of the solidification of SRP waste and storage in a Retrievable Surface Storage Facility (RSSF). This paper reports on the research and development program now in progress at the Savannah River Laboratory.

Characteristics of SRP Wastes

Origins. Most of the radioactive waste at SRP originates in the two separations plants, although some waste is produced in the reactor areas, laboratories, and peripheral installations. The principal processes used in the separations plants have been the Purex and the HM processes, but others have been used to process a variety of fuel and target elements. The Purex process recovers and purifies uranium and plutonium from neutron-irradiated natural uranium. The HM process recovers enriched uranium from uranium–aluminum alloys used as fuel in SRP reactors. Other processes that have been used include: recovery of ^{233}U and thorium (from neutron-irradiated thorium), recovery of ^{237}Np and ^{238}Pu, separation of higher actinide elements from irradiated plutonium, and recovery of enriched uranium from stainless-steel-clad fuel elements from power reactors. Each of these processes produces a characteristic waste.

Most wastes that emerge from these processes are acidic, but are neutralized with sodium hydroxide before transfer to mild steel storage tanks. The waste is first transferred to cooled tanks, where it remains for about two years until most of the short-lived fission products have decayed. The waste is then transferred to uncooled tanks.

After the waste is neutralized and transferred to tanks, insoluble oxides and hydroxides of fission products and of metals (e.g., iron, mercury, and manganese) used in chemical processing settle to form a sludge layer. The volume of the sludge is about 10% of the total waste volume; the remainder (the supernate) consists principally of solutions of sodium salts used in processing or formed when the acidic wastes are neutralized. In recent years, aged supernates have been evaporated and returned to cooled tanks where much of the salt crystallizes. This procedure reduces waste volume and renders the waste less mobile.

Chemical Composition. No completely valid generalizations about the composition of SRP wastes can be made because large variations in the composition of both the sludge and the supernate occur from tank to tank, within each tank, and with time. For present purposes, it is sufficient to list the components present and estimate their maximum and average concentrations. The major chemical components of the waste are shown in Table I together with an estimate of their average concentration calculated on the basis that all of the salt is dissolved, and the resulting

Table I. Chemical Composition of Waste

Constituent	*Average Concentration* (M)	*Constituent*	*Average Concentration* (M)
$NaNO_3$	2.2	NaF	0.02
$NaNO_2$	1.1	Na_3PO_4	0.01
$NaAlO_2$	0.8	NaCl	0.01
NaOH	0.5	KNO_3	0.01
Na_2CO_3	0.3	*$Ca(CO)_3$	0.006
Na_2SO_4	0.3	*HgO	0.002
*$Fe(OH)_3$	0.15	$Mg(OH)_2$	0.001
*MnO_2	0.02	NaI	0.0002

* Compounds in sludge layer.

solution is mixed uniformly with the sludge. The asterisked compounds should exist in the sludge; the remainder in the supernate. Aluminum will exist in both the sludge and the supernate, and the amount in each will depend upon the sodium hydroxide concentration.

The wide diversity of compounds in the sludge is illustrated by Table II, which lists elements that have been detected or are known to be present in the sludge.

Radioactivity and Toxicity. Table III lists the radionuclides in aged waste, their half-lives, the expected radioactivity of each after 10 years aging, and the relative toxicity of each. Relative toxicity is defined as the ratio of the concentration of a given isotope to its maximum permissible concentration in public zone water. Mercury, nitrate, and nitrite (although not radioactive) are also listed because their toxicity is actually greater than that of many of the radioactive nuclides. ^{90}Sr and ^{137}Cs are

Table II. Various Elements in Sludge

Aluminum	Nickel
Barium	Phosphorus
Calcium	Potassium
Cerium	Praseodymium
Chlorine	Promethium
Chromium	Ruthenium
Cobalt	Silicon
Copper	Silver
Iron	Sodium
Lanthanum	Sulfur
Lead	Tin
Magnesium	Titanium
Manganese	Uranium
Mercury	Yttrium
Molybdenum	Zirconium
Neodymium	

Table III. Amounts of Activity in Stored Waste and Relative Toxicity Caused by Radionuclides and Other Components

Isotope	*Half-Life (years)*	*Total Activity*[a] *(curies)*	*Relative Toxicity*[b]
^{90}Sr	28.8	2.0×10^8	2×10^9
^{137}Cs	30.0	2.0×10^8	3×10^7
^{147}Pm	2.6	6.7×10^7	1×10^6
^{106}Ru	1.0	3.5×10^5	1×10^5
^{238}Pu	89	1.7×10^5	1×10^5
^{244}Cm	18.1	1.2×10^5	6×10^4
^{151}Sm	90	4.6×10^6	4×10^4
Hg (inactive)	—	—	2×10^4
^{239}Pu	2.4×10^4	1.7×10^4	1×10^4
$NO_3^- + NO_2^-$ (inactive)	—	—	6×10^3
^{129}I	1.6×10^7	31	1.7×10^3
^{99}Tc	2.1×10^5	3.0×10^4	300
^{79}Se	7×10^4	280	120
^{135}Cs	2.0×10^6	3.1×10^3	100
^{126}Sn	1×10^5	1×10^3	100
^{93}Zr	9.5×10^5	6.7×10^3	30
^{94}Nb	2×10^4	3.2	13
^{107}Pd	7×10^6	26	3
^{158}Tb	150	0.5	4×10^{-5}

[a] Assumed as a basis for this report and does not represent the actual quantities (classified) of aged waste to be processed.
[b] Ratio of concentration in waste to maximum permissible concentration in public zone water.

by far the most hazardous radionuclides. Strontium, plutonium, and the lanthanide and actinide elements predominate in the sludge. Cesium predominates in the supernate, and ruthenium is found in both the sludge and supernate.

Heat Generation. The wastes are self-heating because of the radioactive decay of the fission products. The heating rate of fresh waste exceeds 0.3 W/l. Over 90% of the heat generation is associated with the settled sludge, as nearly all of the fission products except cesium and ruthenium are insoluble in alkaline solutions. The major sources of radioactivity in wastes over 10 years old are ^{137}Cs and ^{90}Sr; each is present at an average concentration of approximately 0.66 Ci/l. (2.5 Ci/gal). The average heat generation rate from these nuclides is 0.0076 W/l. (0.72 Btu/min-ft^3), 60% of which occurs in the sludge.

Criteria for Acceptable Waste Storage Forms

A first step in our program was to define tentative criteria for acceptable waste forms for retrievable surface storage (*1*). The properties generally considered desirable in solid waste forms are shown in Table IV.

Of these properties, high thermal conductivity is not nearly as important for SRP waste as for power reactor waste because of the much lower power density of SRP waste, approximately 0.008 W/l. for SRP waste compared with 100 W/l. for power reactor waste. Even if SRP waste were concentrated by a factor of 100, its power density would be lower.

If the solidified waste is to be water cooled, low leachability is of prime importance in case of a leak in a container. If an air-cooled system is to be used, however, low leachability is only important in case of an accident (floods, explosion, or earthquake followed by rain, etc.).

Mechanical ruggedness is desirable to reduce the probability of the waste forms breaking into small pieces that might be dispersed into the air or rendered more leachable because of the larger surface area exposed.

Minimum volume is important only because it leads to a lower cost. Small volumes may actually be less safe because the greater power density will lead to a smaller thermal inertia (i.e., the waste would heat up more rapidly). The reasons for the other properties shown on Table IV are obvious.

Table IV. Properties Desirable in Waste Forms

High thermal conductance
Low leachability
Chemical and radiation stability
Mechanical ruggedness
Noncorrosiveness to container
Minimum volume
Minimum cost

Evaluation of Potentially Useful Waste Forms

Another step in our program was to survey the literature and evaluate the technology of solid waste forms developed elsewhere (*1*) (Table V). The forms that have been used for solidification of intermediate level as well as high-level waste have been included in the table because the low power density of SRP waste might make these forms applicable.

Calcines can be produced from SRP waste, but were eliminated from consideration because of their high leachability and occurrence as granules, which could make them highly dispersible.

Of the other forms, none seems to be entirely satisfactory or compatible with SRP waste at the present stage of development. Phosphate glass devitrifies and becomes more leachable; borosilicate glass is not miscible with sulfate, forming a highly leachable separate phase; asphalt is flammable, particularly when mixed with manganese dioxide and

Table V. Evaluated Solid Waste Forms

Calcines
Na_2O, Na_2SO_4, $NaAlO_2$, $NaNO_3$

Glasses and ceramics
phosphate
borosilicate
aluminosilicates

Intermediate level forms
asphalt
cement

sodium nitrate; and cement is a highly leachable wast form. In addition, the high temperature processes will lead to some problems with mercury and ruthenium volatilization during processing.

If the sludge and the supernate phases of SRP waste are segregated, and the ^{137}Cs (the only long-lived isotope with a significant biological hazard in the supernate) is removed from the supernate, the sludge and the cesium can more readily be converted to solids of high integrity for transfer to an RSSF. The decontaminated salts can then be dried and stored safely in tanks or bins.

Waste Solidification Studies

Figure 1 indicates various elements of the current waste solidification program at SRP. The purpose of this work is to find a mechanically strong, thermally stable, relatively non-leachable waste form which will retain its desirable properties after receiving a dose of 10^{10} rads. During 1974, all potentially promising waste forms using nonradioactive or tracer-scale simulated waste were under investigation. Evaluation of the promising waste forms takes place in 1975, and the form or forms to be processed will be chosen in 1976.

In the conceptual waste processing plant, supernate will be separated from sludge because the high salt content of supernate would adversely affect any of the promising waste forms. The sludge will then be dried to convert it to a more easily handled form and washed to remove most of the residual soluble salt. The sludge will then be dried again for incorporation into one of these waste forms.

Cesium will be separated from the supernate and loaded onto zeolite. The cesium–zeolite product could then be mixed with a solid matrix, such as cement or glass, to further reduce the cesium leach rate. A possible alternative is to mineralize the cesium by methods developed by Robert Barrer in England (*2*) and by Atlantic Richfield Co. at Hanford

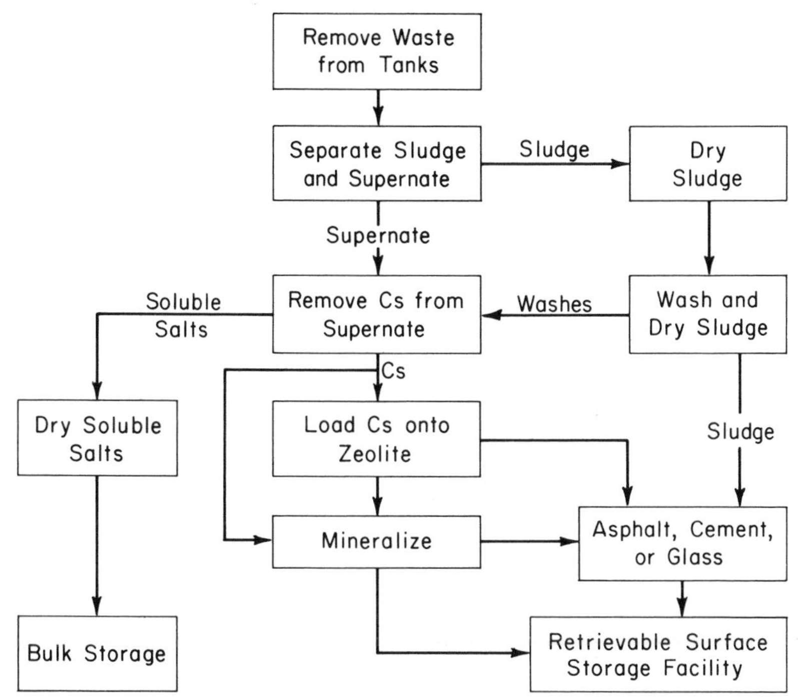

Figure 1. Elements of SRP waste solidification program

(ARHCO) (3). This mineral product might be stored as is or still further protected by solidification in cement or glass.

Removal of Sludge and Tank Decontamination. Waste must first be removed from storage tanks before it can be processed. An engineering study is now in progress to determine removal techniques and the degree to which waste tanks can be decontaminated. Salt is easily removed from the tanks by dissolution in water. However, the sludge is much more difficult to remove, because of limited access through tank openings and in tanks with bottom cooling coils. In the present concept, as shown in Figure 2, the salt is dissolved in water and pumped from the tank. This will be done by pumping diluted supernate solution through rotating nozzles in each tank to slurry and remove the sludge hydraulically. The sludge–supernate slurry is recirculated through the system until no more sludge can be removed hydraulically. It is expected that 98–99% of the sludge can be removed in this manner. Recirculation of the sludge–supernate slurry will require much less water addition than would a once-through water system.

The final 1% or so of sludge remaining will probably have to be removed chemically. Various chemical cleaning solutions are being evalu-

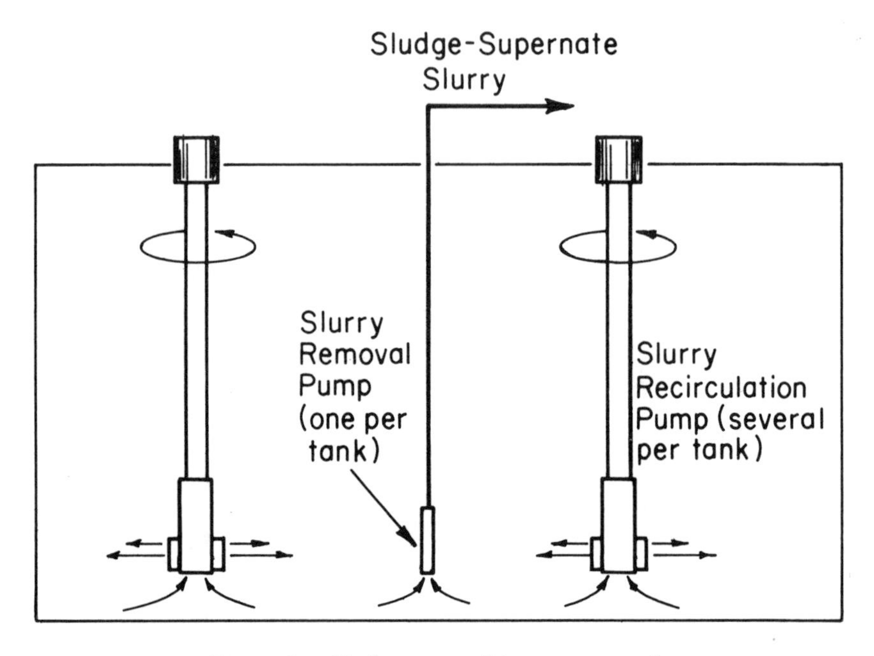

Figure 2. Sludge removal from waste tanks

ated. As these must also be processed to fix the dissolved activity in a solid form, their effect on the final product must also be taken into consideration.

Two assumptions are made concerning the present waste storage tanks: (1) the waste tanks can be decontaminated sufficiently so that they can be decommissioned in place; and (2) activity levels will be reduced to the point that the final stages of decontamination can be done manually.

Sludge–Supernate Separation. After the sludge and supernate are removed from the tanks, the sludge is physically separated from most of the supernate. The clear supernate is sent to ion exchange processing for cesium removal. The sludge then goes through a series of washing and drying steps to remove most of the remaining soluble salts and to minimize the total volume of sludge and the final volume of product.

The first step of this process will be centrifugation of the waste supernate slurry in a slurry centrifuge. Most of the sludge will be removed from the supernate in this step. However, the supernate phase will require further clarification for removal of all sludge fines before the supernate is sent for ion exchange processing. The sludge discharged from the slurry centrifuge will then be dried, washed, and centrifuged at least twice, then dried once more before being mixed with a solidifying agent. The addition of a drying step to precede sludge washing is based on preliminary laboratory tests indicating that subsequent washing and centrifuging are much more effective if the sludge has first been dried.

Cesium Separation. Cesium will be removed from the waste supernate by sorption on a phenol–sulfonic ion exchange resin such as Duolite (Diamond Shamrock Chemical Co.) ARC 359, as shown in Figure 3. This flowsheet is a modification of one currently being used by ARHCO at Hanford (3). Cesium will be absorbed on the two columns in tandem until breakthrough from the first column exceeds a predetermined level, after which the column will be washed with water (not shown in the diagram) and eluted with a mixture of ammonium carbonate and ammonium hydroxide. Breakthrough will be detected by a gamma ray monitor on the line between the two columns.

Experiments with simulated waste indicate that we can process approximately 20 column volumes through one Duolite column and obtain decontamination factors of about 10^4–10^5. We expect this to be sufficient, but, if necessary, we can obtain higher decontamination factors by processing less feed before regeneration or by using a second cycle.

The eluate from this system contains ammonium carbonate, ammonium hydroxide, cesium carbonate, and sodium carbonate. The ammonium salts will be decomposed by heating the solution and collecting the

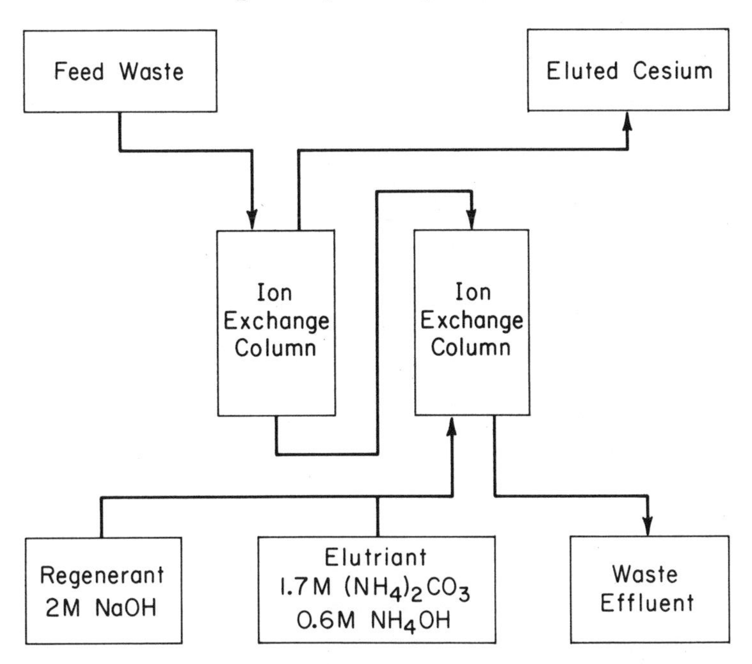

Figure 3. Ion exchange removal of cesium from waste

ammonia and carbon dioxide for recycle. The residual solution will contain only the mixed cesium and sodium carbonates.

The cesium-enriched concentrate will be passed through columns containing a zeolite such as Linde (Union Carbide Corp.) AW–500 (chabazite) where the cesium will be absorbed. Two columns are used in tandem to increase efficiency; when the first column is loaded with cesium, the flow will be diverted to the second column, and a fresh column will replace the first column. The spent sodium carbonate will be recycled back to the feed. The zeolite containing the cesium will go to the solidification plant.

Fixation of Sludge. As mentioned earlier, most of our work to date has been with nonradioactive, simulated sludge. Since sludge composition can vary markedly from tank to tank, we formulated these three types of simulated sludge based on our knowledge of plant processes and some preliminary sludge analyses (Table VI). We expect that these will cover the range of compositions encountered with actual waste.

Early in our work, we found that making small lab-scale batches of simulated sludge was very tedious and time-consuming, and that sludge properties varied from batch to batch. To get around this problem, we made 50-lb batches of simulated sludge in large-scale equipment at our laboratory semiworks. This saved considerable time and ensured that variations in results will not be caused by variations in sludge starting material.

The preparation of "Sludge 2" had been delayed until recently because of a potential health hazard from elemental mercury found in the first few small batches of Sludge 2. Mercurous ion is present to ~0.1% in reagent-grade mercuric nitrate, and mercurous ion reacts with hydroxide to form mercuric ion and elemental mercury. This problem was eliminated by oxidizing mercurous ion with 0.002*M* $KMnO_4$ before adding sodium hydroxide.

CEMENT. Cement is potentially attractive for waste fixation because it is strong, relatively non-leachable when salts are absent, and is made by a low-temperature process with well known technology. We have an extensive program in progress to evaluate all of the common types of cement listed in Table VII. Similar work is also going on at Brookhaven National Laboratory. The five different types of Portland cement contain different proportions of the following compounds: $3CaO \cdot SiO_2$, $2CaO \cdot SiO_2$, $ACaO \cdot Al_2O_3$, and $4CaO \cdot Al_2A_3 \cdot Fe_2O_3$.

Table VI. Simulated SRP Waste Sludges (mole %)

Sludge 1:	50% FeOOH, 50% $Al(OH)_3$
Sludge 2:	40% FeOOH, 40% $Al(OH)_3$, 20% HgO
Sludge 3:	50% FeOOH, 50% MnO_2

All contain 400 ppm Sr

Table VII. Classification of Cements

Portland Cements	
Type I:	normal (general-purpose)
Type II:	modified (lower heat than Type I)
Type III:	high early strength
Type IV:	low heat of hydration
Type V:	sulphate resistant
Portland–Pozzalanic Cement	
Type I–P:	80% Type I, 20% fly ash
High Alumina Cement	
Luminite:	primarily $CaO \cdot Al_2O_3$

Portland–Pozzolanic cement contains a material (in our case fly ash) that reacts with the calcium hydroxide produced by Portland cement to form calcium aluminosilicates to give a cement with improved properties.

High-alumina cement, Lumnite (Universal Atlas Division of U.S. Steel Corp.), is not a Portland cement at all, but consists entirely of calcium aluminate. It reacts with water to form several hydrated species and can give an exceptionally strong and refractory product.

Since virtually nothing was known about incorporating our types of sludge in cement, we decided to study thoroughly the various cement–waste combinations in our nonradioactive work. In our sample preparation program, which is now complete, we made 189 cement–sludge–water formulations. The three levels of cement were 10, 25, and 40 wt % of the cement–sludge total mass. Water, which can be a critical factor, was adjusted to give wet, ideal, and dry consistencies, respectively. In addition, 21 cement–water formulations were prepared as controls.

In the lab, cement and sludge are thoroughly dry-mixed before water is added. The wet paste is poured into 1-in.-diameter x 4-in.-long plastic cylinders and cured in a humid atmosphere for 28 days to develop full strength. Typical samples are shown in Figure 4. At the time of preparation, a portion of the wet paste is transferred to a radioactive hood where a sample is spiked with 10^8 d/m of ^{239}Pu for future leach rate studies.

Our testing program includes measurements of the following properties: compressive strength, leachability, radiation stability, thermal stability (long-term), and impact resistance.

Compressive strength is historically one of the best measures of cement quality and will be useful for design purposes for the RSSF (the Retrievable Surface Storage Facility).

Leachability of a waste form is perhaps the most important criterion in considering accident situations. In the absence of salts, some of our cement products may approach the low leachability of glass. After experimenting with several methods for measuring leach rates, we selected the

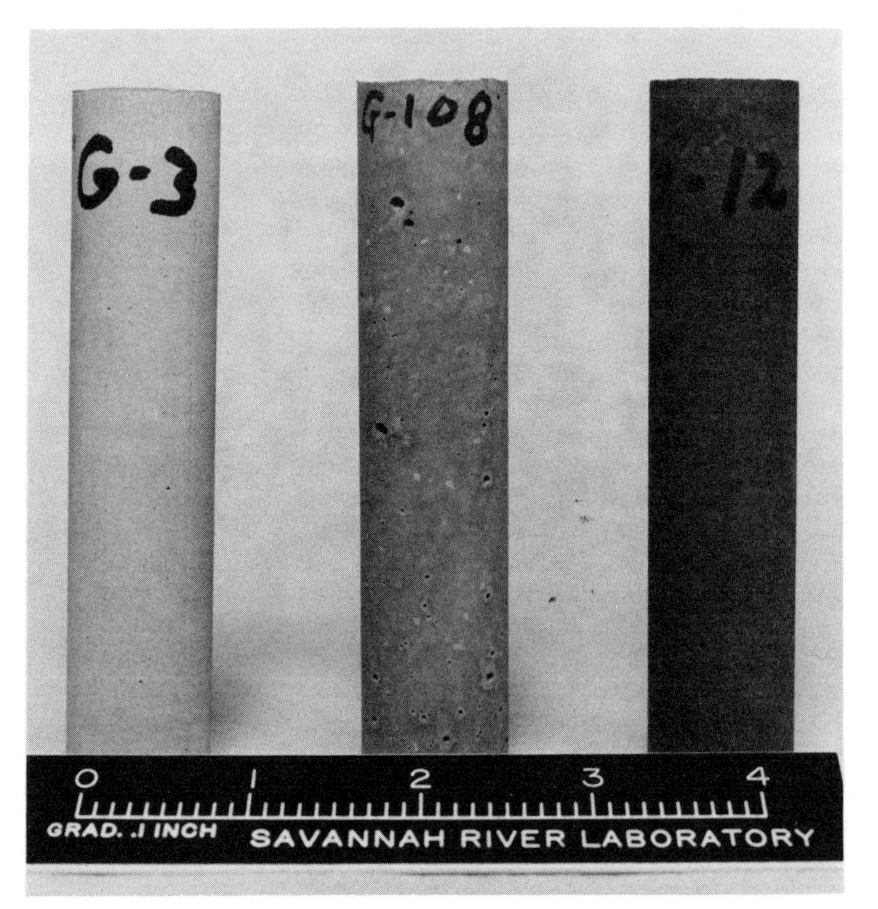

Figure 4. Cement samples

simple static method, and we are now following the standard method proposed by the Leach Rate Subcommittee of the Waste Solidification Demonstration Working Group (WSDWG).

During its lifetime, our waste forms will receive an integrated dose of $\sim 10^{10}$ rad. Forms with initially good properties will be irradiated at 3×10^7 rads/hr with a ^{60}Co source and rechecked for compressive strength and leach rate. A special air-cooled container will be used to keep the sample temperature around 60°C.

Promising forms will be heated for several months at the normal storage temperature and at 100°C to determine possible ill effects. The maximum specific power of cement waste forms would be $\sim$0.4 W/l. with normal storage temperatures around 60°C.

Impact resistance is of concern in accident situations, and tests are now being developed to measure it accurately. Selected samples will be subjected to microstructure analysis, differential thermal analysis, thermogravimetric analysis, and thermal conductivity measurements.

GLASS. We are interested in solidifying sludge in glass because glass has very low leachability, and glass waste forms would have about half the volume of cement forms. The maximum specific power of glass waste forms would be ~0.8 W/l. with normal storage temperature around 60°C. Since some elements do not readily form glasses, we first wanted to demonstrate that our simulated sludges could be solidified in glass, and then to optimize the formulation on the basis of low leach rate, melting point, and high sludge content. The promising formulations will then be studied for radiation stability, release of activity, and release of mercury during their formation and solidification.

We have chosen to work with borosilicate glasses rather than phosphate glasses because of the large increase in leachability of phosphate glasses when they devitrify, as shown in work at Battelle (*4*). The leachability of borosilicate glasses increases only slightly upon devitrification. We have taken advantage of some excellent European work to select glass formulations for our initial studies.

Models for the preparation of good glasses were developed by Grover and Chidley in England (*5*), and by Bocala, Donato, and Sgalambro in Italy (*6*). These models specify the ratios of network-forming elements (silicon, aluminum, and boron) to each other and to other elements, such as oxygen, sodium, calcium, etc., that will yield satisfactory glasses. In addition, we have used a model developed at Hahn–Meitner Institute (*7*) and specific formulations for work at Jülich (*8*) and Karlsruhe (*8*).

In our experiments, the glass components, as either oxides or carbonates, are dry-mixed and heated at 850°C for 1 hr to remove water and carbon dioxide. The mixture is then melted for 1 hr, placed in a 400°C furnace for 1 hr, and then removed to cool to ambient temperature. A typical product is shown in Figure 5. In our initial studies, 14 glass formulations based on these models and the seven formulations shown in Table VIII made acceptable glasses.

All of these glasses were made with the Type 3 Fe–Mn sludge at 1150°C. Mix 5 is based on the English–Italian model. All atom ratios are in the range for a good glass. Mix 6 is based on the Hahn–Meitner formula, Mix 7 is from Jülich, and Mix 8 is the Karlsruhe formula which

Figure 5. Glass samples

Table VIII. Glass Compositions (wt %)

Mix	SiO_2	B_2O_3	Al_2O_3	Na_2O	CaO	TiO_2	*Sludge*
5[a]	50	15	—	10	—	—	25
6[b]	28.8	12.8	12.0	8.0	17.6	—	20
7[c]	38	22	—	8	12	—	20
8[d]	42	8	2	16	4	8	20
9	40	15	10	10	—	—	25
10	30	15	10	10	10	—	25
14	42	8	—	8	14	8	20

[a] "Ideal" composition based on English model.
[b] Hahn–Meitner Institute formulation.
[c] Julich formulation.
[d] Karlsruhe formulation.

contains titanium for cesium retention. Mixes 9 and 10 are variations of Mix 5 with aluminum and calcium substituted for silicon. Mix 14 has the cesium retention properties of Mix 8 and the low melting properties of Mixes 6, 7, and 10.

A summary of the sludge incorporation tests is given in Table IX. In general, a higher temperature is required for the Type 1 sludge because aluminum increases the viscosity of the glass melt. We are currently determining the maximum amount of Type 3 sludge that can go into glass at 1150°C, and we plan to investigate higher temperature melts. Ideally, we would like to operate no higher than about 1150°C because a metal rather than a ceramic melter could be used in the plant facility.

Our program to determine leach rates is just beginning. The data needed are leach rates of plutonium, strontium, and cesium at ambient temperature. The simplest fastest method to calculate leach rates is

Table IX. Summary of Sludge Incorporation Tests

	Quality of Glass				
	Type 3 Sludge (wt %)			*Type 1 Sludge (wt %)*	
Mix	*25*	*35*	*45*	*25*	*35*
5	+[a]	+	+	+	—
6	+	+	+	+	+ @ 1240°C
7	+	+	+	+	+ @ 1240°C
8	+	+	+	i @ 1250°C	—
9	+	+	+	i @ 1250°C	—
10	+	+	+	+ @ 1300°C	—
14		+		+	—

[a] All heated 1 hr at 1150°C; glass homogeneous and pourable; + = good; − = poor; i = inhomogeneous.

measuring weight loss upon leaching at elevated temperature. This method was used in our initial leaching studies.

The glass was crushed, sieved, and leached for 24 hr with boiling distilled water in a Soxhlet-type extractor. Although there are variations, the values reported (Table X) would be reduced two to three orders of magnitude at ambient temperature, and the glasses would then have an acceptably low leach rate. Our program now is to determine the temperature dependence of leach rates and to correlate weight loss data with bulk leach rates based on specific elements.

Table X. Summary of Leach Tests

Weight Loss During Leach, %[a]

	Type 3 Sludge (wt %)			*Type 1 Sludge (wt %)*	
Mix	*25*	*35*	*45*	*25*	*35*
5	0.55	0.49	0.50	0.09	—
6	0.68	0.70	0.65	1.13	1.57
7	0.82	0.99	1.32	1.22	0.59
8	0.31	0.20	0.26	0.21	—
9	0.58	0.45	0.45	0.36	—
10	0.25	0.35	0.28	1.21	—
14		0.30		0.39	

[a] Weight loss of 1% corresponds to leachability of 2.4×10^{-4} g/cm^2-day at 99°–100°C. Soxhlet method, 24 hr, 99°–100°C.

We plan to select two or three of the best formulations and determine whether cesium and ruthenium will volatilize from the melts. Based on some work at Karlsruhe (8), we might not have problems from either element. Cesium apparently reacts with molybdenum trioxide or titanium dioxide to form nonvolatile molybdates or titanates that greatly reduce cesium volatility. We will work with titania because of its greater solubility.

The Karlsruhe workers have also shown that removal of nitrate prior to calcination greatly reduces the volatility of ruthenium as RuO_4. Since our sludge will be washed to remove soluble salts, we think the nitrate ion concentration and thus the ruthenium volatility should be low.

Our remaining problems are to determine the radiation stability and volatilization of mercury from Type 2 sludge. We do not believe it possible to keep mercury in the melt, so our efforts will be concentrated on efficient removal of mercury from the off-gas.

ASPHALT. Our investigation of solidification of sludge in asphalt has been somewhat more limited in scope. We have demonstrated that Types 1 and 3 sludge can be incorporated into asphalt, and we have measured flash points and radiation stability of the products.

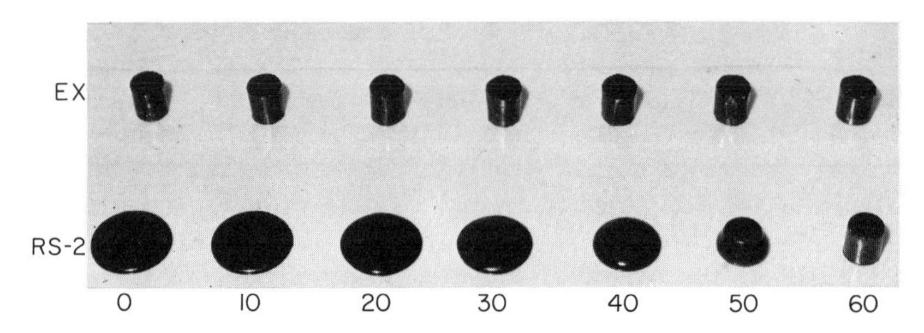

Figure 6. Fe–Al waste in asphalt, wt %

Figure 7. Apparatus for determining ignition points of asphalt formulations

Two types of asphalt were used: "blown" asphalt and emulsified asphalt. "Blown" asphalt is asphalt stock that has been heated at temperatures from 200°–300°C and had air blown through it. The process is one of dehydrogenation and polymerization, resulting in a product that is thermally more stable. We used a blown asphalt provided by the Exxon Corp. because it most closely resembles Mexphalt R 90/40 (Shell Oil Co.), the blown asphalt used most in Europe. Mexphalt R 90/40 is not available in the U.S., but differs chiefly in being somewhat softer than the asphalt from the Exxon Corp.

Emulsified asphalt was also investigated. Type RS–2 was chosen, as this type is common and has been tested at Oak Ridge (*9*). Our material was ~69% asphalt, 28% water, and 3% tall oil. Even after evaporation of the water, the material is still much softer than blown asphalt, as shown in Figure 6.

Blown asphalt is more stable mechanically, but requires higher operating temperatures than emulsified asphalt. Currently, blown asphalt is the favorite for waste fixation purposes in countries using this method.

A significant consideration with asphalt is its flammability. Karlsruhe reports (*10*) that asphalt containing 4% MnO_2 ignited during preparation. We determined ignition points of a wide range of formulation in an apparatus of a Lucite (E. I. du Pont de Nemours & Co., Inc.) box, a vertical tube furnace with a rheostat control, a thermocouple recorder system, and a source of air and nitrogen (Figure 7). This was later modified by positioning the air and nitrogen outlets to blow across the sample and installing a nichrome wire as a high temperature ignition source.

Samples of emulsified or blown asphalt containing 0–60% of Sludges 1, 2, and 3 were placed in beakers and heated individually in the furnace. Samples were heated to 550°C, with most asphalt evaporation and pyrolysis occurring between 300° and 500°C. Samples heated only by the tube furnace did not ignite spontaneously, so an ignition wire was embedded in the sample and energized intermittently during sample heat-up to 200°C; above 200°C, the wire was energized until ignition occurred. Upon ignition, the wire was turned off. Usually flames were self-sustaining only at 300°–500°C. Above 500°C, evaporation and pyrolysis were essentially complete, and remaining vapor was inadequate to support combustion. Nitrogen was used to extinguish flames; ignition in nitrogen was not possible at any temperature.

The flash points for both emulsified and oxidized asphalt are shown in Figure 8 as a function of percent sludge. Values shown are averages of two or three tests, usually reproducible to ±40°C. The slight variation in flash point suggests that the composition had little effect and that asphalt burned independently. This was further demonstrated by additional tests with 5, 10, 20, 30, 40, 50, and 60% sodium nitrate filler and

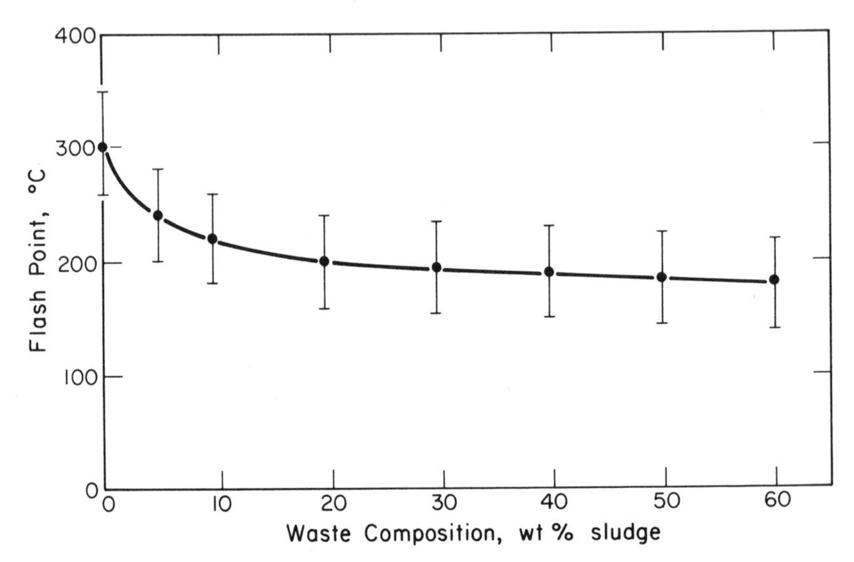

Figure 8. Flash points of emulsified and blown asphalt with Types 1, 2, and 3 simulated sludge waste

with 5, 10, 20, 30, and 40% MnO_2 with 35% nitrate in each. Flash points are within the ranges shown. Once ignition occurred, nitrate-containing waste did burn more rapidly. These results agree well with studies at the Royal Military School at Brussels, which report (*11*) that asphalt–sodium nitrate mixtures are safe providing the temperature does not exceed 280°C.

Flame retardants were not generally effective. The combustion of asphalt occurs in a stable diffusion flame above the material surface. Heat from the flame is transmitted back to the asphalt causing vaporization. Vapors enter the flame, react exothermally, and continue the cycle. Fire retardants must therefore inhibit vapor phase combustion. Retardants tested included ethyl iodide, bromotrichloromethane, methyltrichlorosilane, all of which increased the flash point appreciably, and chloroform, carbon tetrachloride, and potassium bicarbonate, all of which helped only at higher percent sludge formulations.

With the Fe–Mn sludge and asphalt, we also ran a differential thermal analysis (DTA) which showed an exotherm beginning around 100°C. This could be hazardous in a process vessel or in a waste form of large dimensions tightly enclosed in a container.

Samples of blown asphalt containing high amounts of sludge retain their integrity and become very hard when irradiated (Figure 9). All samples made with blown asphalt were irradiated at 6×10^5 rads/hr at 32°C. With 10–20% sludge, the volume increase was ~75%. At 40–60%

sludge, the volume increase was only 0–5%; presumably because the forms are more porous, and radiolytic gases can escape.

The composition of the gases evolved during radiolysis of asphalt–sludge formulations was determined by gas chromatography. Hydrogen is the principal evolved gas, with small amounts of carbon dioxide and methane.

Leach tests and irradiation of some samples to 10^{10} rads are in progress, but we have no data to report at this time. To summarize our work to date, our sludges can be incorporated in asphalt; the products are flammable and release hydrogen upon irradiation.

Fixation of Cesium. The final step in the conceptual cesium purification flowsheet is loading cesium onto Linde AW–500 zeolite, which is essentially chabazite. We will select the best cement formulations from our sludge work to demonstrate the incorporation of cesium–chabazite in cement. In a preliminary test at 1150°C, we made an homogeneous glass with 40 wt % chabazite.

Based on some recent mineralization work, it might be preferable to lock cesium into one of these mineral lattices, depending on the source of feed.

The product from the Duolite columns will be evaporated to form $2M\ Na_2CO_3$–$0.002M\ Cs_2CO_3$. Barrer (2) has found that alkaline solutions of sodium carbonate react with kaolinite to form sodalite containing tightly bound sodium carbonate. We have experiments in progress to determine whether cesium is also bound in the sodalite lattice.

An alternative to formation of sodalite would be formation of cancrinite. Cancrinite and sodalite have the same molecular formula,

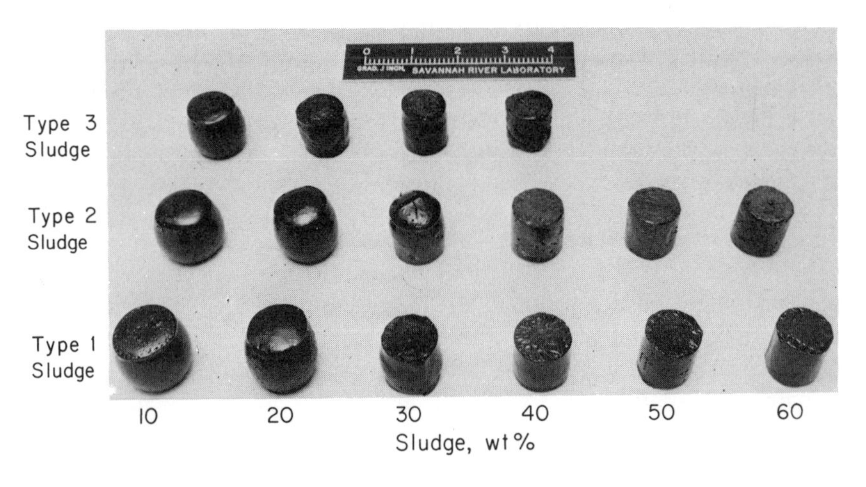

Figure 9. Asphalt–sludge forms after irradiation to 10^9 rads in low-level ^{60}Co cell

differing only in the stacking arrangements of rings of $(Al,Si)O_4$ tetrahedra. Studies at ARHCO (3) have shown that cesium is bound tightly in cancrinite which is formed by reaction of alkaline nitrate solutions with kaolinite. The Duolite columns product could be acidified with nitric acid to remove carbonate and then made alkaline with sodium hydroxide for cancrinite synthesis. The cancrinite product will be prepared for comparison with the sodalite product under identical conditions.

It might be desirable to load cesium onto chabazite before mineralization. Chabazite reacts hydrothermally with aqueous cesium chloride to form pollucite, a naturally occurring mineral with cesium trapped in aluminosilicate cages. Reaction of cesium-loaded chabazite with pure water might produce pollucite or the sodium zeolite, faujasite.

We will test all mineral products for leach rate and radiation stability up to 10^{10} rads, and incorporate them into cement and glass for leach rate studies on the composite product.

Retrievable Surface Storage Facility (RSSF)

After solidification and packaging, our conceptual studies assume that the waste may be placed in an on-site Retrievable Surface Storage Facility (RSSF). For the design basis assumed for this study, there would probably be a total of 9,000–10,000 containers, 2 ft in diameter by 10 ft in length, with an average heat output of $\sim$250 W/container. The RSSF would probably be close-coupled to the solidification plant. The facility is expected to have the following characteristics:

1. The facility should have a lifetime of 100 years or more.
2. The RSSF would probably be a reinforced concrete structure designed to withstand earthquakes, etc.
3. All waste containers would be easily retrievable for repackaging, off-site shipment, etc.
4. Although surveillance and maintenance of the facility would be at a minimum, they would be required for the entire lifetime of the facility.
5. Automatic monitoring systems would be installed to detect any release of activity from waste containers.

Removal of heat from the containers is of prime concern in designing such a facility. Several options are being considered, including: cooled water basin, natural convection in air, and forced convection in air. Particular attention will be given to designing a system that will continue to maintain container temperature at acceptable levels even in the event of postulated accidents or temporary abandonment of the facilities under emergency conditions.

Final design of the RSSF will require detailed cost/benefit studies for the entire system to determine optimum design criteria and to specify engineered safeguards. These studies will consider the form of the waste and the probabilities of all events that could lead to releases from the facility.

Bulk Storage

The decontaminated supernate may be evaporated to a solid salt in a wiped film evaporator. For the design basis assumed in this study, the total volume of solidified salt would be about 100 million l. ($\sim$25 million gal). Since the heat-producing isotopes will be removed, this material can be stored in large containers in bulk storage. It is expected that the decontamination of this material will be sufficient that the principal hazard in the salt cake will be the nitrate itself. The activity left in the salt may well be less than that left in waste tanks after final decontamination. On this basis, further processing of this material into a less leachable form would probably not be necessary. However, it is presently assumed that this material cannot be dispersed to the environment.

The two principal options for the long-term storage of the decontaminated salt cake are: (1) storage in new large containers constructed specifically for the decontaminated salt cake, and (2) storage in existing waste tanks no longer in service. In either case, the salt cake can be removed easily from bulk storage (if necessary) by dissolving the salt in water and pumping it from the containers.

Optimization Studies

Studies are under way to minimize both the costs and risks of retrievable surface storage. Risk is defined as the frequency of an adverse event times the consequences of that event. In the case of radioactive waste storage, the total system risk (sum of the risks of all individual events) is usually measured in units such as radiation dose (man-rem/yr) to the nearby population. However, other non-biomedical factors, such as disruption of operations, adverse public reactions, etc., must also be taken into account for realistic measures of system risk. Techniques, such a fault-tree analysis, are being used to develop quantitative models of total system risk. (Fault-tree analysis (FTA) is a systems safety engineering technique that provides a systematic, descriptive approach to the identification of high risk areas. FTA is a formalized process for identifying the possible modes of occurrence of undesired events in a given system and for estimating their probabilities.) When these quantitative models are developed, they will allow estimation of total system risk as a function of system design. When the costs of the system are also factored in, we will be able to determine the system design which will give the minimum risk for a particular level of total system cost. If this procedure is repeated for several different levels of total system cost, a curve of minimum system risk vs. total system cost can be generated. Final determination of system design must be made by either defining an acceptable level of total system risk or by defining an acceptable ratio for the incremental reduction in risk per additional dollar spent on the system.

Literature Cited

1. Wallace, R. M., Hull, H. L., Bradley, R. F., "Solid Forms for Savannah River Plant High-Level Waste," USAEC Report **DP-1335,** E. I. du Pont de Nemours and Co., Savannah River Laboratory, Aiken, S.C., 1974.
2. Barrer, R. M., Cole, J. F., Sticher, H., "Chemistry of Soil Minerals—Part V. Low Temperature Hydrothermal Transformations of Kaolinite." *J. Chem. Soc.* (1968) (**A**), 2475–2485.
3. Shultz, W. W., Atlantic Richfield Hanford Co., Richland, Wash., private communication, 1973.
4. Mendel, J. E., McElroy, J. L., "Waste Solidification Program, Vol. 10, Evaluation of Solidification Waste Products," USAEC Report **BNWL-1666,** Pacific-Northwest Laboratories, Inc., Richland, Wash., 1972.
5. Grover, J. R., Chidley, B. E., "Glasses Suitable for the Long Term Storage of Fission Products," UKAEA Report **AERE-R-3178,** Harwell, Berks, England, 1960.
6. Bocola, W., Lonato, A., Sgalambro, A., "Survey of the Present State of Studies on the Solidification of Fission Product Solutions in Italy," *Symp. Manage. Radioact. Wastes Fuel Process.,* Paris, France, 1972.
7. Heimerl, W., Heins, H., Kahl, L., Levi, H. W., Lutze, W., Malow, G., Scheiwer, E., Schubert, P., "Research on Glasses for Fission Product Fixation," Report **HMI-B109,** Hahn-Meitner-Inst., Berlin, West Germany, 1971.
8. Rudolph, G., Saidl, J., Drobnik, S., Guber, W., Hild, W., Krause, H., Müller, W., "Lab-Scale R&D Work on Fission Product Solidification by Vitrification and Thermite Processes," Report **KFK 1743,** Kernforschungszentrum Karlsruhe, Karlsruhe, West Germany, 1972.
9. Godbee, H. W., Blanco, R. E., Frederick, E. J., Clark, W. E., Rajan, N. S. S., "Laboratory Development of a Process for the Incorporation of Radioactive Waste Solutions and Slurries in Emulsified Asphalt," USAEC Report **ORNL-4003,** Union Carbide Corp., Oak Ridge National Laboratory, Oak Ridge, Tenn., July, 1967.
10. Bähr, W., Hempelmann, W., Krause, H., Nentwich, O., "Experiences in the Treatment of Low- and Intermediate-Level Radioactive Wastes in the Nuclear Research Centre, Karlsruhe," IAEA Report **SM-137/12,** *Proceedings of a Symposium: Management of Low- and Intermediate-Level Radioactive Wastes,* Aix-en-Province, France, September 7–11, 1970, p. 472.
11. Tits, E., "Investigations on the Hazards Caused by Incorporating Sodium Nitrate and Sodium Nitrite into Bitumen," Eurochemic Report **IDL-67,** Royal Military School, Brussels, Belgium, 1973.

Received November 27, 1974. Work done under Contract AT(07-2)-1 with the U.S. Atomic Energy Commission.

3

High-Level Radioactive Waste Management Program at the National Reactor Testing Station

CYRIL M. SLANSKY

Allied Chemical Corp., Idaho Chemical Programs–Operations Office, National Reactor Testing Station, Idaho Falls, Idaho 83401

A status report is given on high-level radioactive waste management at the irradiated fuel reprocessing plant (ICPP) at the National Reactor Testing Station (Idaho National Engineering Laboratory). Interim storage of high-level liquid waste, calcination of liquid waste by fluidized bed calcination and storage of calcine in bins is summarized over the past decade. Results are given on the conversion of granular calcine to cermet, glass-ceramic, and ceramic forms for alternative long-term storage requirements. The environmental impact of effluents released in the course of waste solidification is given for the ICPP.

Most of the high-level radioactive waste at the National Reactor Testing Station (NRTS) is generated at the Idaho Chemical Processing Plant (ICPP) during the reprocessing of spent nuclear fuel. The fuel comes from nuclear reactors using highly-enriched ^{235}U fuel. The high-level radioactive waste is primarily the first-cycle raffinate from the solvent extraction of dissolved fuel solutions. This waste is self-heating in both the liquid and solid forms. Although the second- and third-cycle solvent extraction wastes are not self-heating, they will be treated as high-level waste. While solid plant waste and trash (contaminated with radionuclides) might be inferred by the title of this paper, such waste will not be discussed extensively here, because the quantity of radionuclides involved is very small compared to those from the first-cycle raffinate.

The principal concern in the management of high-level waste at the ICPP is to store the liquid waste safely within a period not to exceed five years, calcine the liquid waste to a dry solid, and store the solid

Table I. Typical Chemical Compositions of the Liquid Wastes in Interim Storage at ICPP (Molarity, *M*)

Component	*Acid Aluminum*	*Fluoride-Bearing*	*Stainless Steel Sulfate*	*Electrolytic*	*Graphite*[a]	*Second- and Third-Cycle*
H^+	1.0	1.5–2.0	3.2	2.2	2.3	0.1–1.6
Al^{+3}	1.16	0.7	—	0.2	0.3	0.5
Zr^{+4}	—	0.45	—	—	—	T[c]
Cd^{+2}	—	0–0.1	—	—	—	—
Nb^{+5}	—	<0.01	—	—	0.046	—
Na^+	0.065	0.003	—	0.004	—	1.6–2.3
Fe^{+3}	0.005	0.004	0.069	0.09	—	0.02
Cr^{+3}	—	T[c]	0.017	0.03	—	—
Ni^{+2}	—	T[c]	0.01	0.01	—	0.0004–0.003
NH_4^+	—	—	—	0.1	0.06	0.04
Gd^{+3}	—	—	—	0.0025	—	—
Fissium[b]	—	—	—	0.004	—	—
B^{+3}	—	0.2	—	—	0.065	0.005–0.2
Hg^{+2}	0.018	—	—	—	—	0.0007–0.005
K^+	—	—	—	—	—	0.2
NO_3^-	4.6	2.4	2.6	3.2	2.0	4.6–5.0
SO_4^{-2}	0.01	—	0.6	—	—	0.04–0.06
PO_4^{-3}	—	—	—	—	—	0.02
F^-	—	2.5–3.2	—	—	1.2	0.0007–0.007
Cl^-	—	0.001	—	—	—	0.03

[a] Design estimates.
[b] Fissium is composed of 48 w/o ruthenium, 4 w/o rhodium, 4 w/o palladium, and 2 w/o zirconium.
[c] T = Trace, less than 0.1 w/o.

safely in engineered storage bins where it will be retrievable at all times (*1*). The dry storage bins are designed for a lifetime of hundreds of years although removal of the radioactive waste for further treatment or transport to a Federal repository might be accomplished whenever desired.

Sources of High-Level Radwastes at the ICPP

The first-cycle raffinate wastes produced at the ICPP are the acid aluminum waste from various test reactor fuels, fluoride-bearing waste from zirconium-matrix fuel, a small amount of stainless steel sulfate waste from fuel from developmental reactors such as the Organic Moderated Reactor Experiment (OMRE), acid stainless steel nitrate waste from the electrolytic dissolution of Experimental Breeder Reactor–II (EBR–II) reactor fuel, and an acid waste from the recovery of uranium from nuclear rocket fuel (*2, 3*). The second- and third-cycle raffinates for all of these fuels have a common composition. Typical chemical compositions of the liquid wastes are given in Table I. Radiochemical compositions of the contents of three of the first-cycle zirconium-bearing waste tanks are given in Table II and of three tanks containing second- and third-cycle waste in Table III (*4*). Based on the isotopic content for Pu and U and cooling times at the ICPP, the concentrations of ^{241}Am, ^{243}Am, ^{242}Cm, and ^{244}Cm have been calculated for the average high-level zirconium-bearing waste to be 3.8×10^{-5}, 1.1×10^{-6}, 9.9×10^{-10} and 6.6×10^{-8} g/l, respectively. Most of the first-cycle raffinates are self-heating, while the second- and third-cycle wastes and waste-evaporator bottoms

Table II. Radioisotope Analysis of First Cycle Waste (ICPP)

Ci/l

	Tank No.		
Isotope	*WM–185*	*WM–187*	*WM–188*
^{60}Co	1.5×10^{-3}	9.0×10^{-4}	1.5×10^{-3}
^{95}Zr	4.3×10^{-3}	8.7×10^{-3}	$<1.6 \times 10^{-3}$
^{95}Nb	4.9×10^{-3}	1.4×10^{-2}	$<8.9 \times 10^{-4}$
^{106}Ru	1.1×10^{-1}	9.0×10^{-2}	2.5×10^{-2}
^{125}Sb	3.3×10^{-2}	3.0×10^{-2}	1.8×10^{-2}
^{134}Cs	2.1×10^{-1}	2.1×10^{-1}	1.2×10^{-1}
^{135}Cs	1.0	7.0×10^{-1}	8.0×10^{-1}
^{144}Ce	1.0	9.0×10^{-1}	2.0×10^{-1}
^{154}Eu	1.5×10^{-2}	1.1×10^{-2}	1.3×10^{-2}
Gross β	6.5	4.7	3.2
^{90}Sr	9.4×10^{-1}	1.9×10^{-1}	2.9×10^{-1}
^{3}H	8.1×10^{-4}	1.7×10^{-4}	1.2×10^{-4}
	Plutonium Analyses		
^{239}Pu, wt %	72.2	67.1	70.4
^{240}Pu, wt %	17.6	20.5	18.9
^{241}Pu, wt %	7.8	9.2	8.2
^{242}Pu, wt %	2.5	3.4	2.7
Total Pu, g/l[a]	1.93×10^{-3}	1.19×10^{-3}	1.54×10^{-3}
	Uranium Analyses		
^{234}U, wt %	1.7	1.2	1.3
^{235}U, wt %	56.1	72.6	30.8
^{326}U, wt %	5.9	6.3	3.6
^{238}U. wt %	36.3	19.9	64.3
Total U, g/l	4.7×10^{-4}	9.3×10^{-4}	2.8×10^{-3}
	Neptunium		
Np, g/l	1.19×10^{-5}	1.57×10^{-5}	9.74×10^{-6}

[a] ^{238}Pu constitutes an additional 5% to the g Pu/l.

Table III. Radioisotope Analysis of Second- and Third-Cycle Waste (ICPP)

Ci/l

Isotope	Tank No. WM–186	WM–184	WM–180
^{95}Zr	6.2×10^{-5}	1.0×10^{-5}	3.3×10^{-4}
^{95}Nb	3.3×10^{-5}	4.2×10^{-6}	1.0×10^{-4}
^{106}Ru	8.6×10^{-3}	1.5×10^{-4}	3.3×10^{-3}
^{125}Sb	6.7×10^{-4}	2.4×10^{-4}	3.0×10^{-3}
^{134}Cs	1.0×10^{-3}	2.6×10^{-5}	1.0×10^{-3}
^{137}Cs	5.6×10^{-2}	1.2×10^{-2}	2.4×10^{-1}
^{144}Ce	2.8×10^{-3}	4.0×10^{-5}	7.6×10^{-4}
^{154}Eu	2.6×10^{-4}	2.9×10^{-5}	7.0×10^{-4}
^{60}Co	—	1.0×10^{-5}	—
^{90}Sr	4.7×10^{-2}	1.2×10^{-2}	2.1×10^{-1}
^{3}H	6.1×10^{-5}	1.0×10^{-4}	3.4×10^{-5}
Gross β	2.1×10^{-1}	4.4×10^{-2}	7.7×10^{-1}
Plutonium Analyses			
^{239}Pu, wt %	—	88.3	86.1
^{240}Pu, wt %	—	10.3	8.6
^{241}Pu, wt %	—	1.1	4.9
^{242}Pu, wt %	—	0.4	0.4
Total Pu, g/l[b]	[a]	1.893×10^{-3}	0.938×10^{-3}
Uranium Analysis			
^{234}U, wt %	0.09	0.95	0.65
^{235}U, wt %	5.86	64.39	43.70
^{236}U, wt %	0.52	3.07	2.08
^{238}U, wt %	93.53	31.59	53.57
Total U, g/l	0.191	2.01×10^{-2}	2.31×10^{-2}

[a] WM–186 contained a constituent that interfered with the analysis.
[b] ^{238}Pu constitutes an additional 5% to the total Pu.

are low-heat liquid wastes. Intermediate-level wastes are evaporated, the overhead is cleaned by ion exchange columns to less than RCG limits before discharge to the water table, and the bottoms processed as high-level waste. Individual storage tanks generally contain components of several types of waste.

To date the ICPP has produced approximately 4.6 million gal of high-level liquid waste of which 2.0 million gal are in the liquid form as shown in Table IV and 2.6 million have been calcined to a solid. High-level wastes began accumulating at the ICPP in 1953 and the rate of production has steadily increased (3). New wastes are anticipated from new fuels and will be discussed later in this chapter.

Interim Storage of High-Level Radioactive Liquid Wastes

The high-level radioactive liquid wastes are stored in stainless steel 300,000-gal tanks located in underground concrete vaults pending their solidification in the Waste Calcining Facility (WCF). At present the liquid waste is stored for less than four years and is dependent in part on the rate of calcination and rate of waste generation. The tank farm consists of eleven 300,000-gal tanks and four special tanks of 30,000 gal each.

A schematic of one of the 300,000-gal tanks is shown in Figure 1 (*5, 6*).

The 300,000-gal tanks are 50 ft in diameter with a 21-ft high wall and a 32-ft height from the floor to the top of the hemispherical roof. The floor is flat and the sides are vertical. The vessel floors and lower 8 ft of the sides are made from 5/16-in.-thick stainless steel plate, and the hemispherical roof is 3/16 in.-thick stainless steel plate. The tanks are designed for a liquid with a specific gravity of 1.4, for a temperature of 200°F, and a pressure of 3.9 in. water vacuum.

The 300,000-gal waste tanks and the associated concrete secondary containment vaults were analyzed for their dynamic response to a hypothetical earthquake occurring at the NRTS, using the STRAP-D (Struc-

Table IV. Summary of ICPP Tank Farm Liquid Waste Inventory as of February 26, 1974

Tank No.	*Type of Waste*	*Volume (gal)*
WM–180[a]	Second- and third-cycle, and PEW evaporator bottoms	265,000
WM–181[b]	Spare (for service waste diversion)	22,000
WM–182[a]	Fluoride	139,000
WM–183[a]	Nonfluoride	112,800
WM–184[b]	Second- and third-cycle, and PEW bottoms	283,000
WM–185[a]	Fluoride	278,000
WM–186[b]	Second- and third-cycle, and PEW bottoms	283,000
WM–187[a]	Fluoride	277,000
WM–188[a]	Fluoride	14,000
WM–189[a]	Fluoride	144,400
WM–190[a]	Spare	0
WM–103[c]	Stainless steel sulfate	16,600
WM–104[c]	Spare	0
WM–105[c]	Stainless steel sulfate	1,800
WM–106[c]	Stainless steel sulfate	15,000
TOTAL		1,839,000

[a] 300,000-gal tank with cooling coils.
[b] 300,000-gal tank without cooling coils.
[c] 30,000-gal tank with cooling coils.

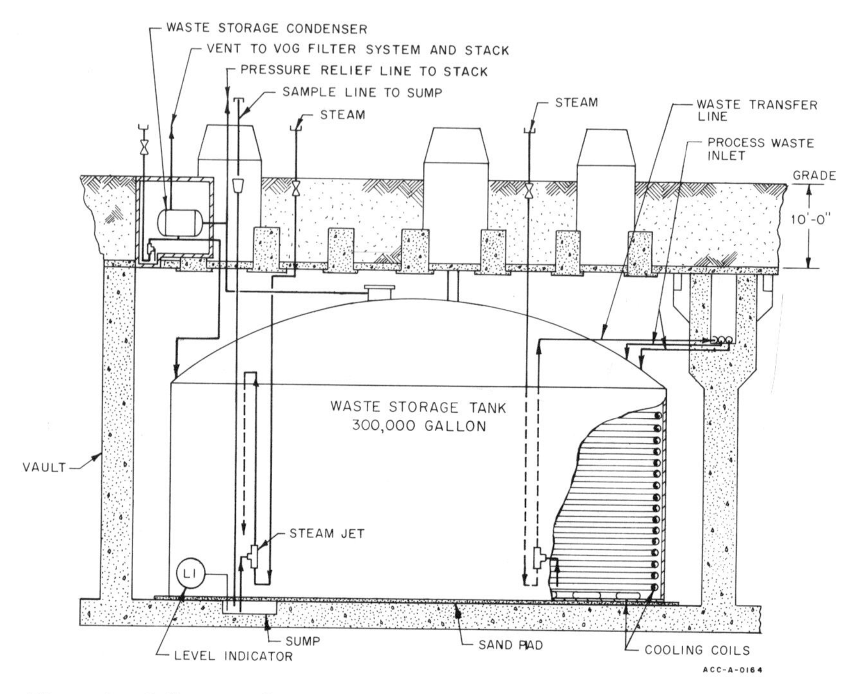

Figure 1. Schematic diagram of typical 300,000-gal liquid waste tank at ICPP

tural Analysis Package–Dynamic) computer code (7). Ground acceleration histories from four of the largest West Coast earthquakes were used as forcing functions for the code—viz, the 1940 El Centro, 1949 Olympia, 1952 Taft, and 1966 Parkfield.

Calculations show that the waste storage tanks and vaults will withstand the severe ground shaking accompanying very strong earthquakes (6). Even if subjected to the 1940 El Centro earthquake record scaled to a peak ground acceleration of 0.5 g, the waste tanks were stressed to 21,300 psi which is 59.1% of the 36,000 psi required to cause tank failure. Using the unscaled 1940 El Centro earthquake record which had a peak ground acceleration of 0.33 g, 50.4% of yield stress in the tanks was produced. An acceleration of 0.33 g corresponds to the acceleration expected at ICPP from a maximum hypothetical earthquake of Richter magnitude 7¾ located on the Arco fault at a point 15 mi from ICPP. It does not take into account the appreciable energy attenuation which likely would occur because of the particular geological structure of the Snake River Plain.

Under the same conditions, analysis of the reinforced concrete vaults housing the storage tanks showed that a peak ground acceleration of 0.5 g would cause a peak stress of 87.7% of yield stress while the unscaled 1940 El Centro record would impose 79% of yield stress of the vaults.

An earthquake producing a 0.33-g peak ground acceleration at ICPP is considered a purely hypothetical upper-limit event which has not been exceeded in magnitude by any earthquake in recent geological times. The Arco fault has not been seismically active in historic time.

The liquid level in the 300,000-gal tanks that contain highly radioactive waste is continuously monitored. A level indicator in each tank is set to alarm when the volume reaches 285,000 gal. This alarm sounds in both the control house and the main process building. A second leak detection system is a liquid-level indicator in the sump of the vault surrounding the tanks.

Internal tank pressure is also measured and transmitted to the control panel. If the relief valve is activated, the pressure is released to a line that vents directly to the stack. The normal tank venting system is connected to the process vessel off-gas system in the main plant, which is a filtered system.

Specific gravity of the liquid waste is measured, and the value is transmitted to a recorder which has an alarm in both the control house and main process building that activates at a specific gravity of 1.35.

Multiple temperature points in the tanks provide individual temperature measurements for each 2 ft of vertical wall height and for the tank bottom. In addition, a complete set of thermocouples is provided on the opposite side of each tank. A total of 24 points is available for each of the cooled tanks (10 for each uncooled tank). The temperature of each tank is controlled below 50°C for corrosion control and is continually recorded on multiple point recorders. A high-temperature alarm is installed for each tank, with alarms at the control house and at the main process building. Normally the wastes are well below the high-temperature alarm point.

Eight of the eleven 300,000-gal tanks contain a closed-loop cooling system consisting of coils of stainless steel pipe around the walls and on the floor of the tank. The water pressure in the cooling coils is greater than the pressure in the waste tanks so that normally if a leak occurred the cooling water would leak into the tank. However, a radiation monitor is installed on the return coolant line from the tank and will alarm in the event the cooling water becomes contaminated. By closing the individual valves on the 30 cooling coils in each tank, and then opening these valves one at a time and observing the radiation monitor, the leaking coil could be located and isolated. Contaminated water can be transferred to the waste evaporator system for processing, and fresh water can be added to the cooling coils. The tanks with cooling coils have been in operation since 1953, and no leak has occurred in any of the tanks or cooling systems to date.

Each of the 300,000-gal storage tanks is contained in an individual concrete vault which serves as a secondary containment barrier if waste leaks from the tank. The vault is constructed of reinforced concrete, approximately 2 ft thick. The vaults are either octagonal or square and are about 60 ft across and 33 ft tall.

The tanks are also connected to reflux condensers, which will condense the vapors if the waste boils. These condensers are stainless steel shell-and-tube condensers contained in concrete housings in the tank farm area. The condenser pits have sumps equipped with jets to transfer leakage back to one of the waste tanks; however, the condensers are not normally used.

Each of the tanks within the vaults rests on a thin layer of sand on top of a concrete pad at the bottom of the vault; the bottom of the vault rests on bedrock. The vaults generally have 1- to 3-ft-deep by 1- or 2-ft sq sumps to collect liquid. An increase in the level in the sump indicates that the tank contents are leaking or groundwater from outside the vault has leaked in. Groundwater has been observed to enter past the cover of the vault which is not sealed. Any liquid in the sumps can be returned either to the tank or to the main plant for processing. Each sump has a measuring device which indicates the liquid level in the sump. The measuring unit is connected to a multipoint recorder that records the liquid level in all of the sumps continuously. The readings for the sumps are recorded in a logbook once each 8-hr shift. The system is also connected to an alarm which sounds in the control house and in the main plant if the level exceeds a predetermined value (15 in.). Water in the vault sumps can be sampled manually by submerging a bottle attached to a handline; remote samplers are being installed.

One of the tanks containing cooling coils always remains empty to serve as a spare; thus, if a leak develops, the contents of the leaking tank can be transferred to the spare. A freeboard volume equivalent to 300,000 gal is also retained in the sum of the other seven cooled tanks so that the liquid from a leaking tank can be distributed among seven tanks instead of all of it going to the spare tank. In effect, two empty tanks are available for any emergency in the tank farm.

Three of the 300,000-gal tanks that do not contain cooling coils are used to store radioactive liquid wastes that do not generate any appreciable heat because the concentration of fission products is too low. One of these three tanks is retained as a spare in the event that failure of a process vessel containing radioactive solution contaminates normally nonradioactive cooling water. If this occurs, the contaminated cooling water is diverted to the spare tank and later processed in the main plant waste system to remove the radionuclides before the water is released to the discharge well.

The four 30,000-gal stainless steel tanks are buried in a horizontal position. They are 12 ft in diameter and have a 38-ft-long straight side. The shell and head of these tanks are 11/16 and 9/16 in. thick, respectively. They also are equipped with cooling coils and are connected to a shell-and-tube condenser to remove heat produced by the decay of radionuclides. These tanks rest on a concrete pad but are not surrounded by a vault. A curb surrounds the concrete pad, and a sump equipped with a level alarm collects any leakage from the tanks. A 24-in.-diameter pipe extends from the surface of the ground to the sump so that a portable unit can be used to empty the sump if necessary.

The four 30,000-gal tanks were used to store wastes from early processes used at the ICPP. One of these tanks has been emptied, one is nearly empty, and the other two are about one-half full but will be emptied in the near future. When emptied, none of these tanks are planned for the routine storage of radioactive liquids, as they do not meet the present secondary containment criterion for storage of high-level radioactive waste.

All underground waste lines (totalling 9,000 ft) in the tank farm are stainless steel. All lines that carry radioactive waste have either: (a) a stainless steel-lined concrete encasement, (b) a secondary steel pipe with centering spacers, or (c) a tile pipe with spacers encased in concrete (*8*). This construction provides two safety features as follows:

(1) It prevents contact of the primary waste line with the soil to avoid underground corrosion.

(2) It provides very rapid detection of leaks in the primary line by the use of monitors.

The lines carrying radioactive waste go through junction boxes constructed of concrete with a stainless steel liner. If radioactive waste leaks into a junction box or the secondary pipe encasement, it drains to a sump in a 300,000-gal tank vault. There is no evidence, direct or indirect, of any leak in the waste lines to date. The jet system on each tank is equipped with a siphon break to avoid creating a siphoning action. During the periods of surface runoff resulting from melting snow or heavy rains, 5 of the 11 vaults containing 300,000-gal waste storage tanks collect water from in-leakage through the top of the vault. This water is periodically transferred into the tank or to the waste evaporator for processing. The path to the vault of the surface water has been traced recently and means to eliminate it are underway.

Figure 2. Aerial view of WCF

Solidification of High-Level Liquid Wastes in the Waste Calcining Facility (WCF)

The policy at the ICPP is to solidify the high-level radioactive liquid wastes and to store on-site the resulting granular solids in engineered bins where the waste will be in safe interim storage and fully retrievable. In the solid form the waste is less mobile than as a liquid; the volume of the waste is reduced by a factor of 9:1; the cost of solidification and storage is less than building more tanks for liquid wastes; and the operation meets the basic philosophy of waste management of the USAEC (now ERDA).

The WCF at the ICPP has been solidifying high-level wastes in a fluidized-bed calciner since December 1963 (*9, 10*). As of January 1974, over 1,510,000 gal of aluminum-type waste and 1,089,000 gal of zirconium-type waste had been calcined to give a total of 42,500 cu ft of solids. The original process was developed for solidifying high-level waste from the processing of aluminum-clad test reactor fuel. The process has been modified to calcine fluoride-bearing zirconium waste, stainless steel sulfate waste, and ammonium nitrate-containing waste. The WCF is a vital unit in an integrated program of spent fuel storage, fuel reprocessing, and waste management at the ICPP. An aerial view of the WCF and three sets of solid storage bins is shown in Figure 2.

A schematic flowsheet of the WCF is shown in Figure 3. Solidification takes place in the 4-ft-diameter by 13-ft-high fluidized-bed calciner. Most of the plant equipment is devoted to cleaning up the entrained radioactive particles in the calciner off-gas. The off-gas passes into a cyclone to remove most of the solids, a quench tank to cool the gas, a venturi scrubber-separator-demister, a silica gel adsorber for the removal of voltaile ruthenium, another cyclone, and finally a high-efficiency HEPA filter to remove submicron particles before discharging to the 250-ft plant stack.

In the calciner, the liquid waste is pneumatically atomized from three nozzles at 85–140 gph into the heated 6-ft-deep fluidized bed of solidified granular waste at 400°–500°C. The process is endothermic; therefore the bed is heated by in-bed combustion of an oxygen-atomized stream of kerosene. The process is started with a bed of granular material such as dolomite, which is replaced by calcine as the operation continues. By preheating the bed with externally-heated fluidizing air to a bed temperature of 360°–400°C, the atomized kerosene will ignite in the presence of a nitrate waste.

Nitric acid and water are condensed in the off-gas scrubbing system to dissolve the entrained calcine. The off-gas scrubber solution is recycled

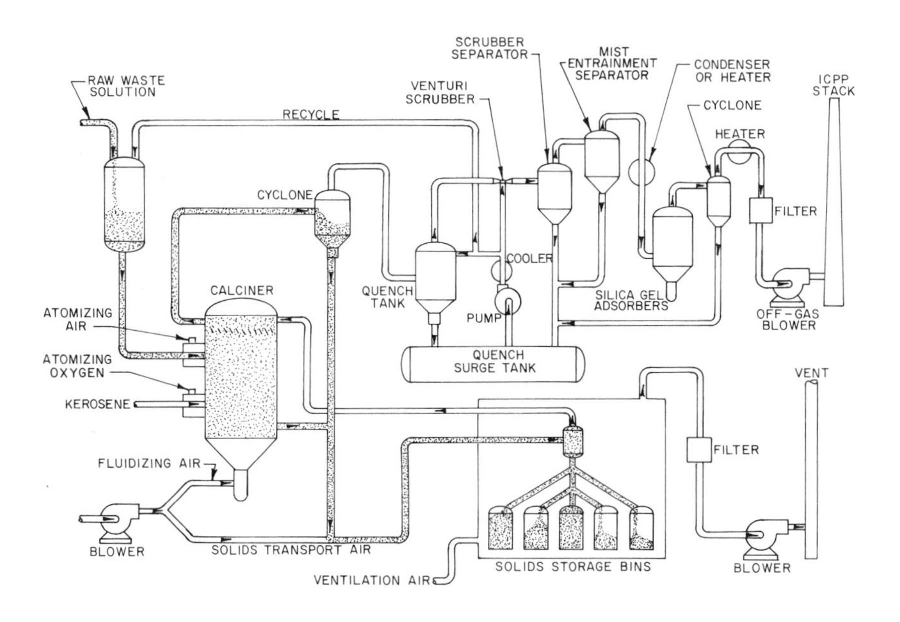

Figure 3. Schematic flowsheet (WCF)

at 15–30 gph by returning it to the calciner feed; the recycle solution is 20–30% of the total feed to the calciner. Based on the cross-sectional area of the empty calciner vessel, the superficial inlet fluidizing velocity is 0.6–1.2 ft/sec. The volumetric ratio of atomizing air to liquid feed is varied as necessary to cause sufficient breakup of bed particles by impingement and thereby control the average bed particle size at a desired value. The value of this volumetric ratio varies from 500–600 for alumina-type waste to a value of 700–800 for zirconium-type waste. Bed height is controlled by the rate of removal of calcined solids.

The calciner originally built was heated by a heat exchanger using liquid sodium–potassium alloy as a transfer medium which was heated externally in an oil-fired furnace. Although 35,000 hr of satisfactory services were obtained with NaK, an in-bed combustion process was developed and installed to obtain a greater processing rate; other changes noted were lower vessel wall temperatures, lower ruthenium volatility, and increased reliability.

Typical physical, chemical, and radiochemical properties of calcined solids from acid aluminum waste and fluoride-bearing zirconium waste are given in Table V. Other components in the calcine, *e.g.*, actinide elements, can be calculated on the basis of the analyses of the liquid waste.

Table V. Typical Properties of Calciner Product

Physical Properties	*Aluminum Waste*	*Zirconium Waste*
Mass median particle diameter, mm	0.56–0.70	0.6–0.8
Bulk density, g/cc	1.0–1.2	1.7
Composition (wt %)		
Zirconium as ZrO_2	—	21.4
Calcium as CaF_2	—	54.2
Aluminum as Al_2O_3	88.2–89.1	21.9
Sodium as Na_2O	1.3–2.0	—
Nitrogen as N_2O_5	3.9–4.1	1.9
Mercury as HgO	2.9	—
Water	2.0	0.6
Gross fission product oxides	0.6	—
Radioactive properties		
Heat generation, Btu/hr-lb	0.3–0.8	0.1–0.2
Principal radioisotopes, Ci/lb		
Strontium-90	4.9–7.2	0.8–2.5
Cesium-134	0.2–0.4	—
Cesium-137	5.2–8.0	1.0–3.0
Cesium-144	1.3–17.3	—
Ruthenium-106	0.04–0.5	0.01–0.03
Zirconium/niobium-95	0–0.5	—
Promethium-147	3.3–10.6	—

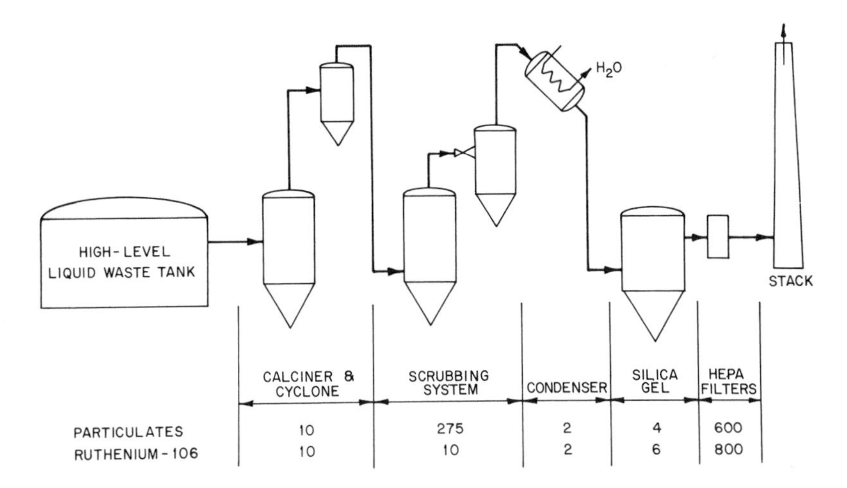

Figure 4. Typical decontamination factors achieved in WCF off-gas system. The decontamination factor for a component is the ratio of the quantity entering to the quantity exiting per unit of time.

Typically current performance of the individual units in the WCF off-gas cleaning system is summarized in Figure 4 (9). Five campaigns have been completed in the WCF. Overall off-gas decontamination factors achieved, when indirect heating was used for the calciner, varied from 0.8×10^7 to 0.9×10^8 for particulates. In campaign 4, in-bed combustion was used and the overall decontamination factor for particulates was 1.4×10^7, and for ruthenium–106 it was 0.8×10^6.

Table VI summarizes typical operating conditions for the WCF as applied to the two major liquid wastes at the ICPP. The conditions for the calciner vessel are based on indirect heating with NaK; in-bed heating gives similar results except that the bed might be subject to short periods of exposure to a high temperature in the immediate vicinity of the burning kerosene. Similarly, the off-gas composition leaving the calciner will depend on the method of heating.

Storage of High-Level Waste Calcine

The calcined product from the WCF is spherical and granular in nature and readily transported pneumatically to the calcine storage bins (100 yds distant) where it will be safe and readily retrievable.

The WCF calcined product is stored in vented stainless steel enclosed bins located in underground concrete vaults (*11*). The original waste storage facility, consisting of four of the annular bins shown in Figure 5a, has been filled with approximately 7,500 cu ft of calcined solids. The

Table VI. Typical Calciner Operating Conditions

	Aluminum Waste	*Zirconium Waste*
Feed Rates		
Gross to calciner, gph	85–100	85–100
Recycle to calciner, gph	15–30	25–30
Calcium nitrate, gph[a]	—	20–30
Net to calciner, gph	60–80	45–55
Product Rate		
To solids storage, lb/hr	50–55	80–90
Calciner Conditions		
Superficial fluidizing velocity, ft/sec	1.0	1.3
Volume ratio of nozzle air to feed rate	515–660	650–800
Bed temperature, °C	400	400
NaK inlet temperature, °C	690–735	690–735
Bed height, in.	73	73
Mass median bed particle diameter, mm	0.5–0.8	0.6–0.8
Scrubbing System		
Solids entering scrubbing system, lb/hr	9–13	10–20
Aluminum, *M*	0.9–1.1	0.2–0.3
Acid, *M*	2–4	2–5
Undissolved solids, g/l	2	1–4
Scrubbing solution rate to quench tower, gpm	60–62	60–62
Pressure drop across quench tower, in. H_2O		3
Scrubbing solution rate to venturi scrubber,	3	
gpm	11–13	11–13
Pressure drop across venturi scrubber, in. H_2O	55–70	55–70
Ruthenium Adsorption and Filtering		
Total gas flow, scfm	900–1600	900–1600
Dew point of gas, °C	59–64	55–65
Adsorber operating temperature, °C	66–74	65–70
Pressure drop across adsorbers, in. H_2O	6–9	6–9
Final filter operating temperature, °C	80–85	80–85
Pressure drop across final filters and adjacent piping, in. H_2O	5–10	5–10

[a] To precipitate fluoride in waste.

second solids storage facility (shown in Figure 5b) consists of seven 12-ft-diameter by 42-ft-high cylindrical bins nested in a 46-ft-diameter concrete vault. The bins are designed for a centerline temperature of 700°C. This facility, having a combined storage capacity of approximately 30,000 cu ft, was filled by February 1972. Thus, construction of a third set of bins (similar to those of Figure 5b) was started during 1969 and filling of the new bins started shortly after their completion in 1972.

During filling of the bins, a cyclone separator above the bins removes the solids from the transport air stream, the solids drop into a vertical standpipe (inside the solids storage vault), and the transport air returns to the calciner for decontamination. Solids collected in the vertical standpipe are directed to one of seven stainless steel bins by inclined lateral lines and a solids flow diverter valve. The lateral lines, five in number, are attached to the standpipe at different elevations, and the corresponding bins fill sequentially as the level of solids rises in the standpipe. A diverter valve located at the base of the standpipe is used to direct the solids, by means of air purges, to either of the two remaining bins. This valve normally blocks the flow of granular solids by use of a horizontal offset which is sealed by the angle of repose of the solid material. When flow into the bin is desired, a small flow of air admitted to the horizontal section disturbs the angle of repose and permits the solids to flow.

Natural-circulation cooling air passes through the vault surrounding the bin and out a stack, thus removing a considerable portion of the heat liberated from the stored solids. Direct radiation from the bins to the surrounding vault wall, and thence conduction into the ground, removes the balance of the heat liberated. The cooling air leaving the bin vault is continuously monitored for radiation (which, if present, would indicate a possible bin failure) and entrainment of particles. Should radiation be detected significantly above background, the monitor sounds an alarm

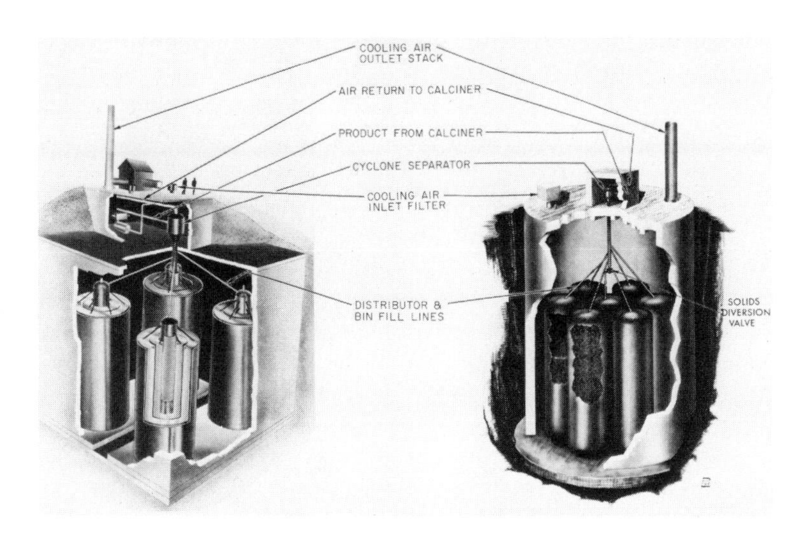

Figure 5a.
First solids storage facility

Figure 5b.
Second & third solids storage facilities

and closes a valve in the cooling air duct. Closing the valve will halt the cooling air flow and prevent contamination of the environment. In a test with a full set of bins with the cooling duct closed, heat was removed through the vault wall with only an increase of 50°C in the centerline temperature in the bins. Each vault contains a sump, sample lines, and jet with liquid-level instrumentation for detecting the entry of groundwater, rainwater, or floodwater into the vault. If liquid is detected, it can be transferred to the ICPP liquid waste evaporator system. No liquid has been detected in the vaults since initial operation of the WCF.

Special features of the storage facility include installation of corrosion coupons and provisions for solids removal. Some 160 sample coupons of various alloy metals—ASTM Types 405, 304, and 304L stainless steels, and AISI Type 1025 carbon steel—have been attached to cables and inserted into the bins. At various periods extending up to 80 years, the samples will be withdrawn, examined, and measured for corrosion. The results should assist in determining the best construction materials for such long-term solids storage. A set of corrosion samples was removed in 1973 from a bin of alumina waste and another set from zirconia waste. No particle agglomeration was evident, nothing was coated on the metal specimens, and the corrosion rate was minimal (*12*). Provision for a solids removal system have been made, since it is possible that advancing technology or integral failure of the bins in the facility may make it desirable or necessary to remove the calcined solids. A 6-in.-diameter pipe extends from the top of each bin upward through the vault roof. If required, the pipe can be cut off and a pneumatic solids removal system inserted down the pipe for retrieval of the solids from the bin.

Laboratory-scale tests, performed on radioactive solids during the first WCF processing campaign, established the feasibility of using a 700°C maximum temperature for design of the second set of storage bins. These initial tests have now been extended to determine high temperature effects on radioactive solids for periods lasting at least one year. The significant conclusions of this work are: (1) essentially no cesium volatilizes at temperatures up to 700°C, but significant amounts volatilize at temperatures of 800°C or higher; (2) no strontium or cerium volatilize at temperatures of up to 1200°C; (3) the volatilization of appreciable quantities of ruthenium and mercury is highly time-dependent, but does occur at temperatures as low as 525°C (ruthenium volatilizes very slowly, while mercury volatilization is rapid); (4) the originally amorphous alumina formed in the WCF will convert to either gamma or alpha crystalline alumina during storage, depending on temperature (only gamma alumina forms below 800°C; alpha alumina predominates above this temperature); and (5) volatile constituents condensed in cooler regions and, in absence of gas purges, were completely retained within the solids. On the basis of these results, it is believed that no radionuclides can escape from stored solids at high temperatures; instead, movement of the heat sources due to radionuclide migration may result in a stable situation in which the maximum temperature of the solids is lower than that without migration.

The solids storage facilities for WCF product have been designed and constructed to withstand any foreseeable natural or man-made occurrence without extensive damage or release of radioactivity to the environment. The storage vaults and bins were designed to withstand an earthquake of magnitude 7 with an epicenter as close as 20 mi. In addition, the stainless steel bins have sufficient corrosion allowance that their integrity can be maintained for at least 500 years, even if the storage area were totally flooded because of some unexpected change in present climatic conditions.

Post Calcination Treatment of Radioactive Calcine

Although WCF calcine is stored at the ICPP in a safe condition and fully retrievable, no decision has been made at this time to ship it to a federal repository. In the event that the calcine were to be moved, criteria for the shipping container or of the form of the solid waste have not been fully established. It is possible that one might be able to ship and store the calcine as it exists. On the other hand, one might prefer to change some of the characteristics of the granular calcine in order to reduce dispersability or leachability.

A development program is underway at the ICPP to examine methods of treating WCF calcine; some methods retain the granular nature of the WCF product while others convert it to a more massive form. Methods of treatment include: (1) conversion to a cermet by the incorporation of the calcine in a metal matrix such as aluminum or lead-tin alloy; (2) conversion to a ceramic by the addition of a ceramic material such as clay and sintering; (3) incorpoartion into a concrete by the addition of binders or cement; and (4) coating of the calcine particles with pyrolytic carbon, glazes, metals, etc.

Of course, the conversion of WCF calcine to a borosilicate glass has been demonstrated as a possibility since 1963, but this phase of the study is not being performed by Allied Chemical since extensive work on incorporation of waste in glass is being studied at Battelle's Pacific Northwest Laboratories.

The experimental program is in the early stages and a comparison of the products of various methods of treatment is only preliminary. A listing of advantages and disadvantages depends on the criteria to be imposed on the product as well as on little-known variables such as cost.

Table VII. Qualitative Evaluation of

Property	*WCF Calcine*
Thermal conductivity (cal/sec/cm²/°C/cm)	~.0007
F. P. leach rate (g/cm²/day)	50–95% Cs 30% Sr in 7 weeks
Temperature stability (°C)	~700°
Radiation stability (R) gross α,β,γ dose	$>10^{12}$
Ruggedness	Fair
Volume increase over calcine	
Retrievability for future improvement	Very good
Uncontrolled storage safety	Fair

However, a tabulation of properties of typical WCF calcine incorporated by various post treatment methods is given in Table VII (*13*). Also shown are the properties of the untreated WCF calcine. In some instances the numerical values of the property vary by several orders of magnitude, while in others only qualitative terms are used. The term retrievability is very important and can mean either a simple recovery (e.g., as from the WCF storage bins) in preparation for packaging and shipping to a repository, or recovery and use or conversion to another form. For example, if one converts a calcined waste to a glass or ceramic block, the unit may be easily retrievable for shipment but may be very difficult to process to another form or to recover specified fission products. On the other hand, untreated WCF calcine has had a minimum of treatment, can be converted to any other form, can be readily processed for desired components, and is fully retrievable.

Environmental Impact of Waste Solidification at the ICPP

The direct impact of waste management at the ICPP on a segment of the general population is minimized by siting the plant at the NRTS, a remote government reservation of 900 sq mi. Releases of minor quantities of radionuclides under routine operating conditions are monitored by continuous air and water sampling and measuring instruments. Potential releases of radionuclides under postulated accident conditions and their environmental impact have been calculated and reported in various safety review and environment impact documents (*7, 10, 11, 14*).

Routine Releases. The high-level waste operations under consideration here are the storage of liquid waste, waste calcination, and storage of calcine in bins. In the 20 years of storing high-level liquid waste, there have been no instances of releases to the ground, nor even of leaks of waste from the tanks to their surrounding vaults. Similarly, there have

the Post Treatment of WCF Calcine

WCF Calcine Post Treatment Method				
Cement	*Cermet*	*Ceramic*	*Graphite*	*Glasses*
.0007–.005	.05–.5	.0007–.005	.005–.01	.002–.004
$\sim 10^{-4}$	10^{-5}–10^{-7}	10^{-5}–10^{-8}	10^{-4}–10^{-7}	10^{-5}–10^{-7}
200°	300°–1200°	600°–1500°	600°–1500°	500°–900°
$>10^{12}$	$\sim 10^{14}$	$\sim 10^{14}$	$>10^{12}>$	$>10^{12}$
				May devitrify
Good	Very good	Good	Fair–Good	Good
20–60%	20–100%	20–80%	30–150%	20–50%
Poor	Fair–Good	Poor–Fair	Poor–Fair	Poor
Fair	Good	Good	Fair	Fair

been no releases of waste from the solid calcine storage bins and vaults. Decay heat from the liquid storage tanks and the calcine bins is discharged to the ground and atmosphere. The temperature rise of the ground in the immediate vicinity of these vaults is a few degrees and has no significant impact on man's environment.

The WCF calciner generates a significant quantity of fine calcined particles which are removed from the gas stream in the air cleaning equipment with a decontamination factor of approximately 1×10^6. Volatile compounds of radioruthenium and tritium are also formed; the ruthenium is removed to low levels in adsorbers and the tritium is released to the environment. Table VIII gives the average annual releases to the atmosphere of tritium, ^{106}Ru, ^{90}Sr, ^{137}Cs, and total alpha from the WCF at an operating rate of 2100/gal/day for 150 days/yr); the average concentration at the NRTS boundary for these radionuclides assuming the nearest downwind boundary and a dispersion factor, X/Q of 3.8×10^{-8} s/m^3; and the Radiological Concentration Guide values for these isotopes in air for a 168-hr week in an uncontrolled area (*14*).

From a comparison of the RCG_a with NRTS boundary concentra-

Table VIII. Releases of Radionuclides to the Atmosphere from the Normal Operation of the WCF

Nuclide	*Ci/yr to Atmosphere*	*Av. Conc. (μCi/ml) at NRTS Boundary*	*RCG_a (μCi/ml)*
Tritium	1200	1.4×10^{-12}	2×10^{-7}
^{106}Ru	1	1.2×10^{-15}	2×10^{-10}
^{90}Sr	1	1.2×10^{-15}	3×10^{-11}
^{137}Cs	1	1.2×10^{-15}	5×10^{-10}
Total alpha	4×10^{-3}	4.8×10^{-18}	6×10^{-14}

Table IX. Atmospheric Releases of Chemical Pollutants from the WCF

Chemical	*Release Rate (T/d)*	*Av. Conc. at NRTS Boundary ($\mu g/m^3$)*	*Air Quality Std ($\mu g/m^3$)*
NO_x	1.6	0.26	100
CO	1	0.2	10,000
Particulates	1.5×10^{-9}	2.5×10^{-10}	75

tions, the atmospheric releases have had a very small impact on the environment. It should be pointed out that the daily release rates vary by as much as a factor of 20 and the values in the table are annual integrated releases.

Liquid wastes are generated in the decontamination of the WCF during down times for maintenance. The PEW wastes are evaporated, the bottoms recycled, and overhead further cleaned in an ion exchange column. The final effluent is discharged to the service well at 50,000 gal/yr of low-level liquid waste as condensate. These wastes are also discharged to the service waste system without significant effect on the environment.

In addition to gaseous and liquid waste, solid waste contaminated with radionuclides are produced by the WCF. Routine wastes of this nature include contaminated rags, blotting paper, plastic wrap, gloves, shoe covers, wood, glass, floor sweepings, scrap metal, HEPA filters and prefilters, spent ion exchange resin, and used equipment. Most of these wastes are packed in cardboard boxes and disposed of in the NRTS Burial Ground. The HEPA filters from the WCF off-gas system are buried at the Burial Ground in their stainless steel housings. In the future it is planned to leach the great majority of radionuclides from the spent HEPA filters and recycle the waste to the calciner. The annual rate of generation of the solid waste is 500 cu ft with a content of approximately 40–100 curies of gross fission product activity.

Nonradioactive wastes are released during calcination; the major components are nitrogen oxides, carbon monoxide, and particulates. Estimated releases are shown in Table IX for WCF operation at 2,100 gal/day for 150 days/yr and an atmospheric dispersion factor (X/Q) of 3.8×10^{-8} s/m³. Air quality standards shown in this table are from 42CFR410.

Actual and Postulated Abnormal Releases. A few relatively minor incidents, resulting in releases of radionuclides, have occurred in the 10 years of operation of the WCF. As already stated, there have been no releases from the liquid storage tanks or the calcine bins.

Incidents are described briefly:

a. In January 1972, approximately 1 Ci of particulate activity—principally ^{106}Ru—was released from the ICPP during WCF operation. The activity was released as very fine particles, but the cause was never found.

b. In May 1972, an open decontamination line in the calcine transport air line resulted in the release to the ground surface of approximately 10 Ci of long-lived fission product activity. The majority of the release was packaged with accompanying soil and sent to the NRTS Burial Ground; the remainder was covered with clean soil and left in situ.

An analysis of land commitment, water resource commitment, impact on biota and impact on human activity has been made on the basis of the general radiation levels at the NRTS and vicinity in 1972 (*14*). The contribution of the WCF to the total activity at the NRTS indicates negligible environmental impact to the general public (*15, 17*).

The main concern in the management of high-level waste is the environmental impact of a postulated maximum credible accident. The major environmental effect from an accident would be from the storage of high-level liquid waste; accidents to the WCF or to the solid storage bins will have an appreciably lesser effect. However, the entire high-level waste system is designed to resist earthquakes, tornadoes, variations in temperature, and floods of maximum credible magnitude.

Solidification of Other Wastes at ICPP (*16*). Aluminum nitrate and zirconium fluoride containing high-level liquid waste have been routinely calcined at the ICPP. Wastes containing ammonium nitrate and also stainless steel sulfate wastes have been satisfactorily calcined in the WCF. Calcination process flowsheets are being developed for new wastes from fuels to be processed at the ICPP such as from graphite fuels and the High Temperature Gas-Cooled Reactor (HTGR). A flow sheet remains to be demonstrated for the million gallons of 2nd and 3rd cycle raffinates and PEW evaporator concentrate which has accumulated (*see* Table IV). The latter is a high-level but low-heat producing waste. The major problem in calcining this waste is its high Na^+ content which results in agglomeration of the bed by $NaNO_3$. The same problem is experienced in calcining neutralized Purex waste. Additives to form high-melting sodium compounds show promise in maintaining fluidization of the bed. The stainless steel nitrate waste from electrolytic dissolution of EBR fuel is blended with aluminum waste for liquid storage, and based on pilot-plant studies the blend can be readily calcined in the WCF.

Pilot-plant work has been performed for the Savannah River Laboratory and for Battelle's PNL to develop fluidized-bed process flowsheets for solidifying neutralized AEC waste and for acid Purex waste. A Purex waste, partially neutralized and containing $2M$ Na^+ (type PW–6), will require an additive to prevent agglomeration in the fluidized bed, a problem already referred to in the calcination of ICPP second- and third-cycle waste, and one which is encountered with neutralized AEC waste containing 6*M* Na^+. A number of additives have been tested in the laboratory and pilot plant, but a satisfactory flowsheet has not been demonstrated. The acid Purex waste studies have direct application to waste from commercial fuel reprocessing plants. In fact, results on simulated acid Purex waste (type PW–4b) indicate satisfactory operation in a 6-in. diameter pilot-plant fluidized-bed calciner. The properties of the solid calcine are summarized in Table X.

Table X. Properties of Calcined Simulated PW–4b Waste

Volume reduction factor under continuous operation	30
Bulk density of bed	2.2 g/cc
Bulk density of fines	1.2 g/cc
Product-to-fines ratio	2.5
Mass median particle diameter	0.30 mm
Attrition resistance index	20–30
Structure (x-ray diffraction)	amorphous

Despite the excellent production record of the WCF, it is showing evidence of wear, and replacement is inevitable. The anticipated fuel reprocessing rates, combined with a limited capacity to store the resulting high-level liquid waste, will require greater processing rates than presently available from the WCF. The corrosion of equipment—due largely to fluorides for which the WCF was not designed—has led to greater deposition of radionuclides, particularly ^{106}Ru, which cannot be sufficiently decontaminated to avoid radiation exposures to maintenance personnel commensurate with current safety regulations. Consequently, a New Waste Calcining Facility has been proposed for the ICPP for installation by the end of 1978. The net throughput will be approximately 3,000 gal/day. The plant will embody the major features of the WCF with upgrading of materials of construction, increased remote maintenance capability, and improved cleanup of the process off-gas.

Conclusions

High-level radioactive waste is produced at the ICPP during the recovery of spent highly-enriched nuclear fuels. The management of the high-level waste is performed in accordance with the latest requiremnts and regulations for such waste. Liquid waste is stored safety in doubly-contained tanks made of stainless steel. The liquid waste is calcined to a solid and stored safely in a retrievable form in doubly-contained underground bins. The calcine can be treated further or left untreated in anticipation of ultimate storage. Fluidized-bed calcination has been applied to many kinds of high-level waste. The environmental impact of high-level waste management at the ICPP has been negligible and should continue to be negligible.

Literature Cited

1. Pittman, F. K., "Plan for the Management of AEC Radioactive Waste," USAEC Report **WASH–1202,** 1973.
2. Offutt, G. E., Cole, H. S., "Run Report for the First Campaign of Coprocessing Fuels at ICPP," USAEC Report **IN–1472,** June 1972.
3. Wheeler, B. R., Dickey, B. R., Lohsee, G. E., Black, D. E., Rhodes, D. W., Buckham, J. A., "Storage of Radioactive Solids in Undeground Facility: Current ICPP Practices and Future Concepts," *Symposium on the Disposal of Radioactive Wastes into the Ground,* Vienna: International Atomic Energy Agency, 1967, pp. 421–440.
4. Rhodes, D. W., Allied Chemical Corporation, ICPP, private communication, August 1972.
5. Holcomb, W. F., Lakey, L. T., Lohse, G. E., "Management of Radioactive Wastes at the Idaho Chemical Processing Plant," 66th Annual Meeting of the American Institute of Chemical Engineers, Nov 11–15, 1973.
6. Hogg, G. W., Holcomb, W. F., Lakey, L. T., Jones, L. H., Coward, D. D., "A Survey of NRTS Waste Management Practices," USAEC Report **ICP–1042,** Sept 1971.
7. "Calcined Solids Storage Additions—Environmental Statement," USAEC Report **WASH–1529,** April 1973.
8. Paige, B. E., Seidenstrang, F. A., Niccum, M. R., "Evaluation of Hazards and Corrosion of Buried Waste Lines in NRTS Soils," USAEC Report **ICP–1913,** Sept 1972.
9. Lakey, L. T., Wheeler, B. R., "Solidification of High-Level Radioactive Wastes at the Idaho Chemical Processing Plant," *Proceedings of the Symposium on the Management of Radioactive Wastes from Fuel Reprocessing,* Paris, France, OECD/AEN and IAEA, March 1973.
10. Lakey, L. T., Bower, J. R., Jr., "ICPP Waste Calcining Facility Safety Analysis Report," USAEC Report **IDO–14620,** 1963.
11. Lohse, G. E., "Safety Analysis Report for the ICPP High-Level Solid Radioactive Waste Storage Facility," USAEC Report **ICP–1005,** Dec 1971.
12. Hoffman, T. L., Allied Chemical Corporation, ICPP, private communication, Oct 1973.
13. Berreth, J. R., Allied Chemical Corporation, ICPP, private communication, Oct 1973.
14. Schindler, R. E., Allied Chemical Corporation, ICPP, private communication, Nov 1973.
15. Markham, O. D., "Environmental and Radiological Monitoring at the National Reactor Testing Station During FY-1973," Paper at Symposium "Ecological Impacts of Nuclear Facilities," AIChE, Philadelphia, November 14, 1973.
16. Dickey, B. R., Allied Chemical Corporation, ICPP, private communication, Aug 1973.
17. Robertson, J. B., Schoen, R., Barraclough, J. T., "The Influence of Liquid Waste Disposal on the Geochemistry of Water at the National Reactor Testing Station, Idaho, 1952–1970," USAEC Report **IDO–22053,** Feb 1974.

RECEIVED November 27, 1974.

4

Solidification and Storage of Hanford's High-Level Radioactive Liquid Wastes

W. W. SCHULZ and M. J. KUPFER

Chemical Technology Laboratory, Research and Engineering,
Atlantic Richfield Hanford Co., Richland, Wash. 99352

When the present Hanford Waste Management program is completed in 1980, approximately 140 million liters of solid salt cake, produced by evaporation of aged radioactive liquid waste will be stored in underground mild steel tanks. Additionally, megacuries of ^{90}Sr and ^{137}Cs will have been removed from Hanford waste solutions, converted to solid $^{90}SrF_2$ and $^{137}CsCl$, and stored underwater in doubly encapsulated metallic containers. Several alternative modes for long-term storage/disposal of these high-level wastes are presently being evaluated. For some of these modes conversion of the solids to immobile silicates of low water solubility may be desirable. Laboratory-scale studies of both low and high temperature processes for preparation of silicate minerals and glasses are in progress; status of the work is reviewed.

Radioactive wastes have accumulated at Hanford since 1944 when the first reactor fuel was processed for plutonium recovery. High-level liquid wastes generated by the Purex, Redox, and $BiPO_4$ processes have been stored as neutralized slurries in 151 underground storage tanks.

In 1957 a program was undertaken and has progressed through the years to insure safe containment of the high-level waste. This effort has developed into the current Hanford Waste Management Program whose chief goal is to assure isolation of hazardous radioisotopes from life forms. Commencing in 1968 the Hanford Waste Management Program has been operated to remove and package as $^{90}SrF_2$ and $^{137}CsCl$ the long-lived heat emitters, ^{90}Sr and ^{137}Cs, from both stored and currently generated high-level wastes. The liquid waste remaining after removal of ^{90}Sr and ^{137}Cs is made alkaline, if necessary, and returned to the underground tanks for eventual evaporation to a solid salt cake. This salt cake is principally a mixture of $NaNO_3$, Na_2CO_3, $NaAlO_2$, and NaOH containing small amounts of ^{90}Sr, ^{137}Cs, ^{239}Pu, and various other radioisotopes. When the current Hanford Waste Management Program is eventually completed some 140 million l. of this salt cake will be stored in the underground tanks and megacuries of $^{90}SrF_2$ and $^{137}CsCl$, doubly encapsulated in appropriate metallic containers, will be stored in water basins.

Some additional description of the objectives and implementation of the present Hanford Waste Management plan is presented in the first part of this review paper. The discussion is deliberately restricted, however, since this technology has been adequately reviewed in detail by various speakers and authors (*1–5*). Recent papers by Lenneman (*2*), Smith (*4*), and Larson (*5*) are particularly informative. A bibliography of publicly available literature concerning radioactive waste management at Hanford has also just lately been published (*6*).

This paper mentions the various modes now being considered for long-term storage of salt cake and encapsulated fission products. Processes to increase immobilization of the stored high-level waste are currently being developed; status of the development work on these processes is also reviewed.

Solidification of High-Level Liquid Wastes

Classification and Source of High-Level Wastes. Hanford high-level wastes can be broadly categorized, as Smith has done (*4*), into high-heat wastes and low-heat wastes. The former consists primarily of Purex and Redox process solvent extraction wastes from fuels processing (Table I).

Table I. Hanford High-Heat Wastes (Concentrations in Molarity)

	Current Acid Wastes		*Stored Neutralized Wastes*	
Constituent	*PAW* [a]	*ZAW* [b]	*Purex*	*Redox*
H	1.0	1.0	–	–
Na	0.7	0.4	3.5	7.0
Fe	0.25	0.26	0.35	–
Al	0.25	0.35	–	–
Zr	–	0.03	–	–
AlO_2	–	–	0.2	1.2
NO_2–NO_3	2.4	2.1	3.1	3.6
CO_3	–	–	–	0.1
SO_4	0.4	0.6	0.5	0.03
PO_4	0.01	0.01	0.01	–
OH	–	–	pH 9–11	pH 7.0
F	0.005	0.1	0.004	–
SiO_3	0.1	0.01	0.1	–

[a] From Al-clad fuel.
[b] From Zr-clad fuel.

Table II. Hanford Low-Heat Wastes (Concentrations in Molarity)

	Neutralized		*Coating Waste*		*Purex Organic Wash*
Constituent	*$BiPO_4$*	*TBP*	*Al-Clad*	*Zr-Clad*	*Waste*
Na	7.0	8.0	5.0	3.0	0.15
Fe	0.1	0.06	–	–	–
Bi	0.03	–	–	–	–
Zr	–	–	–	0.4	–
AlO_2	–	–	2.0	–	–
MnO_2	–	–	–	–	0.02
NO_2–NO_3	5.3	6.6	1.7	1.7	0.1
CO_3	–	–	–	–	0.06
SO_4	0.16	0.31	0.03	–	–
PO_4	0.9	0.29	–	–	–
OH	–	0.18	1.2	–	0.03
F	0	0	0	2.9	–

This material contains greater than 99% of the fission products from the irradiated uranium fuels. It frequently contains enough decay heat to self-boil until sufficient quantities of the short-lived fission products have decayed to more stable isotopes. Then the action of natural heat removal mechanisms keeps the waste solution temperatures below the boiling point. The Redox plant was shut down in 1966 and the last Purex plant campaign was completed in 1972 so that presently no high-heat wastes are being generated. The Purex plant is currently scheduled to start up at least once more to process stored N Reactor fuel.

Low-heat wastes contain relatively small quantities of fission products and associated decay heat. These wastes (Table II) consist of stored $BiPO_4$ process and early Redox process wastes, tributyl phosphate process wastes from an early uranium recovery program, process solvent wash wastes, and fuel cladding removal wastes.

Table III. Current Hanford High-Level Liquid Waste Management Action Plan

High-Heat Wastes
- Remove long-lived heat emitters ^{90}Sr and ^{137}Cs.
- Encapsulate ^{90}Sr (as SrF_2) and ^{137}Cs (as CsCl) for long-term storage on site.
- Treat residual salts as low-heat wastes.

Low-Heat Wastes
- Vacuum evaporation—solidify to "salt cake."
- Remove or sorb free liquid.
- Isolate tanks.

Present Waste Management Program. PROGRAM OVERVIEW. The action plan currently being followed for the management of Hanford high-level liquid wastes is listed in Table III. This plan was adopted after intensive review of the goals and bases of the overall Hanford Waste Management Program.

Figure 1 is a flow diagram for the Hanford high-level liquid radioactive waste management program. Currently sludges and alkaline supernatant liquid from self-boiling wastes stored in underground tanks are being processed in a former separations plant (B Plant) to remove and store ^{137}Cs and ^{90}Sr as nitrate solutions. ["Sludges" are the solids (e.g., $Fe_2O_3 \cdot xH_2O$) which precipitated when the original acid wastes were made alkaline; they contain substantial amounts of ^{90}Sr and plutonium.] Subsequently the resulting wastes are solidified by evaporation and crystallization of the residual bulk salts in the in-tank solidification system. Removal of ^{137}Cs and ^{90}Sr from the high-heat wastes prior to solidification is necessary to prevent abnormally high temperatures in the salt cakes. Although none are presently being produced, self-boiling acid wastes generated in Purex process operation are treated similarly except that after removal of ^{90}Sr and ^{137}Cs a three- to four-year aging period is required to permit short-lived fission products to decay before the residual salt waste can be solidified. Low-heat, nonboiling wastes are

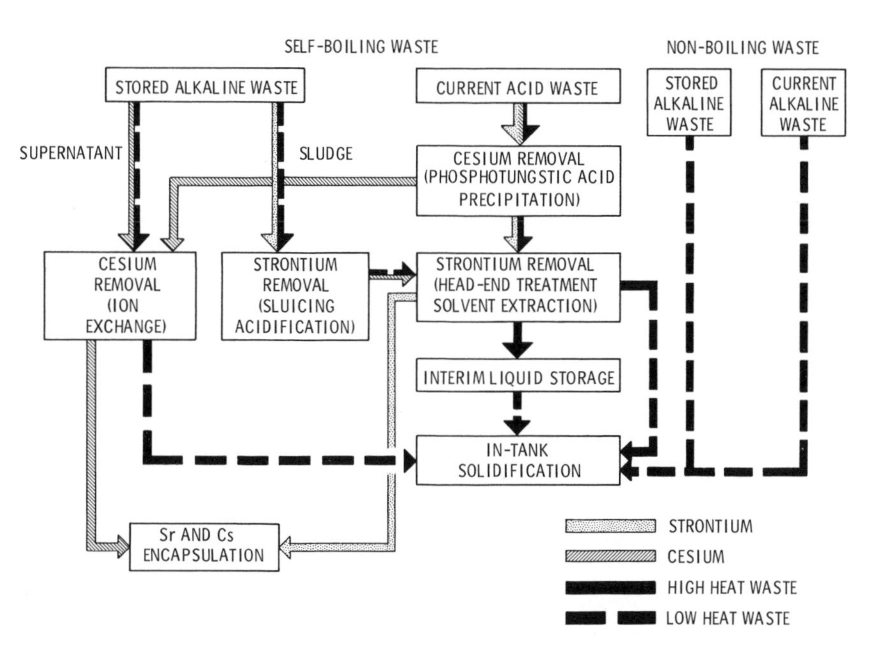

Figure 1. Flow diagram for Hanford high-level liquid waste management program

Figure 2. Salt cake in underground storage tank

routed directly to the in-tank solidification systems for processing to salt cake.

Figure 2 is a photograph of salt cake laid down in one of the underground tanks. The estimated inventory and characteristics of the salt cake to be stored in the waste tanks by about 1980 are listed in Table IV. The composition range shown for salt cake is for air-dried material. In reality salt cake will likely contain varying amounts of interstitial recycle liquor and should be more properly termed "damp" or "wet" salt cake.

Storage of Solidified Wastes

Long-Term Storage Alternatives. How and where should the salt cake and fission product capsules generated in the present Hanford Waste Management Program be stored to ensure positive protection of the public and environment over the long time the radioactivity in these

Table IV. Anticipated (1980) Salt Cake Inventory and Composition

Volume: 140×10^6 liters
Tanks: 73

Typical Composition Range (wt%)	
$NaNO_3$	70–100
$NaNO_2$	2–10
NaOH	0–5
$NaAlO_2$	0–5
Na_2CO_3	2–10
Other	
Fe, SO_4, PO_4, etc.	1

Typical Radionuclides, Ci/l.	
^{137}Cs	0.04
^{90}Sr	0.004
^{239}Pu	$< 10^{-4}$

wastes will be biologically hazardous? Much effort is currently being expended by both the Energy Research and Development Administration (ERDA) and Atlantic Richfield Hanford Co. (ARHCO) personnel to provide a satisfactory answer to this question.

Currently, four alternative modes for long-term storage of salt cake and encapsulated ^{137}Cs and ^{90}Sr are being evaluated (Table V). The first of these is simply to leave the solid wastes in the underground tanks and water basins where they will be when the present Hanford Waste Management Program is completed. The second alternative provides for adding engineered improvements (e.g., support to tank walls, additional

Table V. Long-Term Storage Alternatives for Hanford High-Level Wastes

Order of Prefer-ence	*Storage Mode*	*Waste Product Form*
I	PRESENT	
	Present tanks	Salt cake
	Water basins	Encapsulated $^{137}CsCl$ & $^{90}SrF_2$
II	ENGINEERED IMPROVE-MENTS	
	Present tanks	Salt cake
	Water basins	Encapsulated $^{137}CsCl$ & $^{90}SrF_2$
III	ONSITE REPOSITORY	Immobile silicates
IV	OFFSITE REPOSITORY	Immobile silicates

concrete covers, etc.) to the underground tanks. The two latter alternatives both involve removal of salt cake and, possibly, encapsulated fission products to designated repositories either at Hanford or offsite. Both of these latter alternatives involve conversion of the solid wastes to a more immobile, nonleachable silicate mineral or glass.

Increased Immobilization. Both a low-temperature aqueous silicate process and high-temperature endothermic and exothermic processes are being developed in our laboratories for conversion of salt cake to silicate minerals and glasses. These immobilization processes may find application if a decision is made against continued storage of salt cake in the underground tanks either with or without ancillary engineered improvements. In this event retrieval of the salt cake and storage in a specially constructed repository at Hanford (Alternative 3, Table V) would become a viable storage mode. For this storage mode conversion of the waste to immobile silicates and aluminosilicates of low-water leachability prior to storage may be desirable.

AQUEOUS SILICATE PROCESS. In the Aqueous Silicate process powdered aluminum silicate clays such as kaolin or bentonite are reacted with alkaline solutions or slurries of radioactive waste at temperatures in the range of 30°–100°C. The reaction produces the mineral cancrinite, a sodium aluminosilicate with the radioactive isotopes trapped in the aluminosilicate framework. Electron photomicrographs show the product to be a mass of round clusters of cancrinite crystals approximately 0.5 μm in diameter. To improve the hardness of the product of the Aqueous Silicate process, it may be desirable either to add a binding agent (e.g., CaO, $SiO_2 \cdot H_2O$, tetraethyl orthosilicate, etc.) to the mixture of waste solution and clay or to bond the cancrinite crystals, after water washing, with cement or some organic polymer. A third alternative for producing a suitably hard product is to fire air-dried bricks of clay-waste paste at $\simeq$ 800°C; such treatment converts cancrinite to nepheline.

Aqueous silicate product forms with widely varying properties such as leach rate, hardness, and volume can be produced at various costs by using the three alternative process routes. Typical leach rates for the products in water range from 10^{-5}–10^{-2} g/cm^2-day. In general, the volume of a product from the Aqueous Silicate process is greater than the volume of the original waste. A more in-depth description of the application of the Aqueous Silicate process to immobilization of various high-level radioactive wastes is provided in another paper presented at this symposium (7).

HIGH-TEMPERATURE PROCESSES. *Silicate Melt Process.* One of the high-temperature processes we are studying involves simple melting of a mixture of salt cake with either crushed basalt rock and B_2O_3 or sand

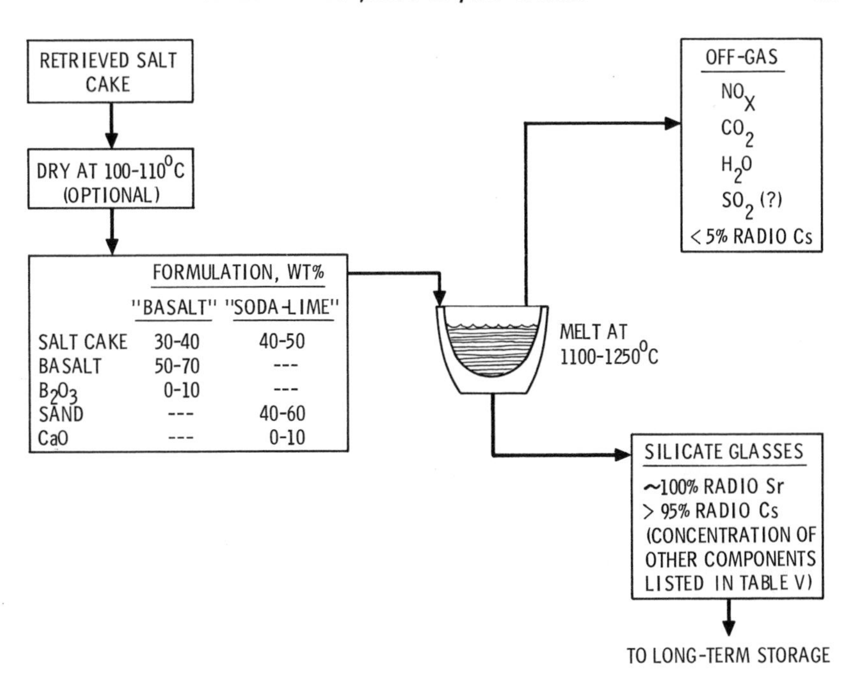

Figure 3. Conceptual flowsheet for application of the silicate melt process to salt cake

and CaO to produce a dense silicate glass. We call this glass-making scheme the Silicate Melt process.

Our batch-scale experiments with both synthetic and actual salt cake of the composition indicated in Table IV have culminated in the conceptual glass-making scheme indicated in Figure 3. This process envisions retrieval of the bulk of the salt cake by dry-mining techniques. Following preliminary drying, if necessary, the retrieved salt cake is converted to glass using either the "soda-lime" or the "basalt" formulation.

The basalt formulation employs Columbia River-type basalt as a source of silica. Extensive deposits of this basalt underlie the Hanford Reservation so that it is readily available as a raw material for large-scale glass-making. The chemical composition of typical Columbia River basalt is shown in Table VI; physical properties of this type of basalt have been determined by Krupka (*8*) and Leibowitz, Williams, and Chasanov (*9*). We also note that incorporation of radioactive waste material into melted basalt was studied briefly earlier in Czechoslovakia by Saidl and Rálkóva (*10*).

Batch tests, described in detail in Ref. *11*, of the basalt formulation with both simulated and actual salt cake show that satisfactorily immobile

Table VI. Hanford Basalt—Typical Composition

Component	*Wt %*
SiO_2	52
$FeO \cdot Fe_2O_3$	14
Al_2O_3	13
CaO	8
MgO	4
Na_2O	3
TiO_2	2.5

glasses are obtained when the process charge contains 30–40 wt % salt cake; leach rates of glasses made from charges containing more than about 40 wt % salt cake are inordinately high. Addition of B_2O_3 to the basalt-salt cake mixture is beneficial, not only to lower the melting range from about 1100°–1150° to 1000°–1050°C but also to reduce volatilization of radiocesium to 5% or less of that in the salt cake. The volume of the glass obtained according to the flowsheet conditions of Figure 3 is about equal to the volume of the salt cake in the original charge.

Using the soda-lime formulation (Figure 3), charges containing as much as 50 wt % salt cake can be successfully converted to what are judged acceptable glasses for long-term storage. The volume of such glasses is 10–24% less than the volume of salt cake in the original charge. Thus, costs of facilities and containers for safe, long-term storage of soda-lime glasses would be less than for glasses made with the basalt formulation. Substitution of cheap CaO for the more expensive B_2O_3 also provides additional cost savings. The principal disadvantage of the soda-lime formulation is that its use requires melter equipment capable of operation in the 1200°–1300°C range rather than in the lower 1000°–1100°C range required with the basalt-B_2O_3 formulation.

The off-gas from melting of mixtures of salt cake with either basalt and B_2O_3 or sand and lime consists mainly of NO_x and CO_2 from decomposition of nitrate, nitrite, and carbonate salts. The off-gas will also contain water vapor and, possibly, traces of SO_2. Not unexpectedly, some radio-cesium also volatilizes when salt cake is converted to glass at 1100°–1200°C. Our laboratory-scale tests suggest, however, that of the order of 5%, or less, of the radio-cesium will volatilize when salt cake is converted to glass in a conventional continuous glass melter.

For example, only 1.3–4.7% of the ^{137}Cs volatilized when 100- to 150-g charges containing either 30 wt % salt cake–60 wt % basalt–10 wt % B_2O_3 or 50 wt % salt cake–40 wt % sand–10 wt % B_2O_3 were heated to 1100°C over a 0.5- to 3.5-hr period. From 6.3–7.6% of the ^{137}Cs volatilized when charges containing 50 wt % salt cake–40 wt % sand–10 wt % lime were heated to 1200°C over a 0.5- to 3.5-hr period. The lower volatility of cesium from charges containing B_2O_3 is in accord with results obtained in our earlier studies (*11*); Mukerji and Kanyal (*12*) state that a solid-state reaction between $CsNO_3$ and B_2O_3 occurs at temperatures as low as 130°–150°C. Other workers (*13*) have also reported that titania in glass mixes acts to decrease volatilization of ^{137}Cs. In our tests addition of small amounts of TiO_2 (5 wt %) to the soda-lime formulation decreased its melting point from about 1200°C to about 1000°C but did not decrease volatilization of cesium.

The average compositions of glasses made from Hanford salt cake are listed in Table VII. Also shown in Table VII is the average composition (*14*) of nonradioactive, commercially produced soda-lime glass. The composition of the soda-lime-type glass made from salt cake is quite similar to that of the commercial product; the radioactive glass contains slightly more Na_2O and slightly less silica than currently produced commercial soda-lime glass. Large quantities of soda-lime glass are made and marketed each year throughout the world; its properties have been characterized in great detail. Availability of this great fund of knowledge provides yet another reason for selecting a soda-lime formulation for making glass from salt cake.

Salt cake glasses made using the basalt formulation contain nearly the same amount of Na_2O as do the soda-lime-type but only about half as much silica. The decreased silica content is compensated for by increased amounts of the network formers Al_2O_3 and Fe_2O_3. At least part of the iron in Columbia River basalt is present as FeO, and at least some FeO will persist even in the molten state at 1000°–1100°C. That FeO strongly absorbs energy in the infrared region is well known (*15*). This

Table VII. Salt Cake Glass Compositions

	Composition, wt %		
	Salt Cake Glasses[a]		
Component	*Basalt Formulation*[b]	*Soda-Lime Formulation*[c]	*Commercial Soda-Lime Glass*[d]
SiO_2	37	60	70 –75
Na_2O	22	26	12 –18
B_2O_3	13	–	
Al_2O_3	10	1	0.5– 2.5
$FeO \cdot Fe_2O_3$	10	–	
CaO	6	13	5 –14
MgO	2	–	0 – 4

[a] Composition of minor components (e.g., PO_4, SO_4, TiO_2, trace metal oxides, etc.) not shown.
[b] 35 wt % salt cake–55 wt % basalt–10 wt % B_2O_3.
[c] 45 wt % salt cake–45 wt % sand–10 wt % CaO.
[d] Composition range listed in (*14*).

property of FeO causes difficulties in continuous glass melter operation in that it lowers thermal conductivity of the melt and causes undesirable thermal gradients to be set up.

Various physical properties of glasses made from both actual and synthetic salt cake using both the basalt and soda-lime formulations have been measured. Typical values of some of these properties are listed in Table VIII. Microstructural and compositional characteristics of representative specimens of the nonradioactive glasses were studied by J. L. Daniels of Battelle Pacific Northwest Laboratories using both conventional microscopy and microprobe methods. Daniels found both the basalt-type and soda-lime-type glasses to have a highly uniform microstructure and composition. The average pore size ($\sim 2\ \mu m$) of the soda-lime glass was somewhat smaller than that in the basalt glass.

Table VIII. Hanford Salt Cake Glasses—Typical Properties

	Typical Values[a]	
	Basalt-Type Glass[b]	*Soda-Lime-Type Glass*[c]
Color—appearance	Green-black; obsidian-like	Transparent window glass
Density, g/cm^3	2.6–2.8	2.4–2.5
Leach rate,[d] g/cm^2-day	6×10^{-5} [e]–2×10^{-7} [f]	3×10^{-5} [e]–2×10^{-6} [g]
Viscosity, poise	75–17[h]	127–66[i]
Thermal conductivity at 23°C, watts/meter °C	0.65	–

[a] For glasses made from both synthetic and actual salt cakes.
[b] For glasses made from charges containing 30–40 wt % salt cake, 50–65 wt % basalt, and 5–10 wt % B_2O_3.
[c] For glasses made from charges containing 40–50 wt % salt cake, 40–55 wt % sand, and 5–10 wt % CaO.
[d] Based on ^{137}Cs; glass products leached in deionized water at 25°C.
[e] Initial 24-hr leach period.
[f] Leach rate after 461 days' leaching.
[g] Leach rate after 100 days' leaching.
[h] At 1100°–1200°C.
[i] At 1250°–1350°C.

Techniques used to determine leach rates of powdered glasses in deionized water at 25°C with a Paige-type apparatus have been described in detail previously (*11*). Initial (24-hr) leach rates, based on ^{137}Cs, of both the basalt-type and soda-lime-type glasses range, typically, from 5×10^{-5} to 1×10^{-6} g/cm^2-day. Upon further exposure to deionized water, leach rates decrease, in a well-known manner, 10–15 times to values in the range 2×10^{-6} to 10^{-7} g/cm^2-day. Glasses exhibiting such leach behavior are judged acceptable vehicles for long-term storage of Hanford salt cake.

Because of their low radionuclide content, the center line temperature of glasses made from salt cake is expected to be in the range 25°–30°C. These glasses are not expected to devitrify upon long storage. Even so, knowledge of time-temperature conditions where these glasses do devitrify and the consequence of crystallization upon leach behavior is of interest. We have heated representative samples of nonradioactive salt cake "basalt" and "soda-lime type glasses" at both 500° and 700°C. The soda-lime type glass did not devitrify when heated two months at either temperature; leach rates of heated glass specimens were within a factor of three of those for unheated material. The basalt-type glass also did not devitrify when heated two months at 500°C but did devitrify when heated a month at 700°C. The leach rate of the resulting crystallized material was 100-fold higher than that of the glass-form.

The Silicate Melt process appears highly amenable to continuous large-scale operation, not only with salt cake, but also with various other Hanford high-level solid wastes including $^{137}CsCl$, $^{90}SrF_2$, sludges, and contaminated soil. Conditions and results of typical process tests with inert CsCl and SrF_2, unwashed Redox process sludge, and various contaminated soils are summarized in Table IX. The soils used in these tests were contaminated either in recent well-publicized leaks or in previous ground-disposal of low-level plutonium wastes. In addition to producing glasses of low leachability, these tests showed that Hanford soils can be substituted on a one-for-one basis for basalt in the Silicate Melt process charge. Addition of Na_2CO_3 to charges containing SrF_2 and/or soil is beneficial to provide a fluxing agent.

Exothermic Process. Various investigators both in the United States (*16, 17*) and abroad (*18, 19*) have proposed the use of thermite-type processes for solidification of Purex process high-level waste. Our version of such a process for immobilizing salt cake is called the Exothermic process. A conceptual flowsheet for the Exothermic process is shown in Figure 4. Feed to the process is prepared by mixing the salt cake with appropriate amounts of Fe_2O_3 (or MnO_2), silicon (or aluminum) metal, and Hanford sand (Hanford sand contains typically 64% SiO_2, 12.7% Al_2O_3, 6.4% FeO, 4.1% CaO, 3.2% Na_2O, 2.0% MgO, 1.8% K_2O, and 1.0% TiO_2). When ignited, these ingredients react exothermically to yield a glassy iron silicate in which other silicates and aluminosilicates, including those of various fission products, are dispersed. With silicon metal, the principal reaction in the Exothermic process is

$$4Fe_2O_3 + 3Si \rightarrow 3Fe_2SiO_4 + 2Fe; \Delta H = -245 \text{ kcal}$$

which serves as a source of heat energy to melt the feed ingredients. The typical silicate product is a green-to-black colored glass whose leach rate

Table IX. Properties of Glasses

Waste Type	Charge, wt %			
	Waste	*Basalt*	*B_2O_3*	*Na_2CO_3*
[a]Sludge: Redox-type	30	60	10	–
[c]Pu-contaminated soil	80	–	10	10
Inert CsCl	20	70	10	–
Inert SrF_2	20	70	–	10

[a] Unwashed sludge contained (all wt %): 25.3 NO_3, 19.1 Na, 7.7 Al, 6.8 Fe, 5.2 SO_4, 4.2 CO_3, 2.7 NO_2, 1.7 Cr, 1.2 Si, 1.0 OH, and 0.5 Ni. Sludge also contained (all μCi/gram) 12000 ^{90}Sr, 549 ^{137}Cs, 116 ^{125}Sb, and 43 ^{154}Eu, as well as 0.04 mg Pu per gram.

[b] Leach rate (based on ^{137}Cs) after 435 days' leaching.

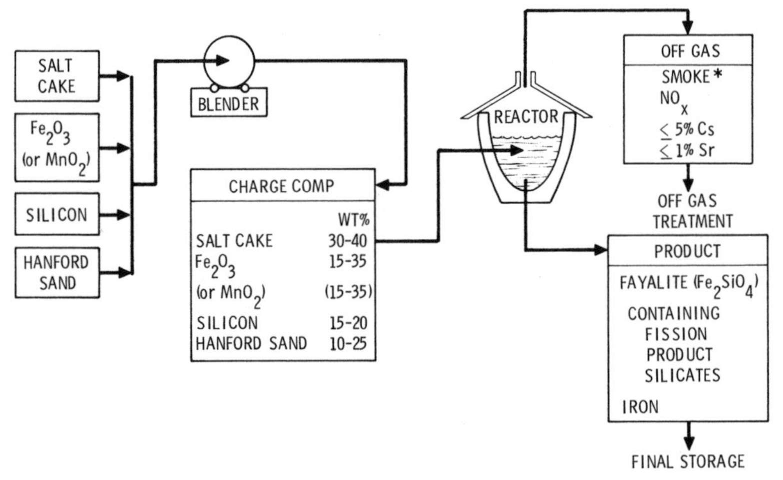

Figure 4. Conceptual flowsheet for application of the exothermic process to immobilization of salt cake

in water is very low—10^{-5} to about 10^{-7} g/cm²-day. Charges containing up to 40 wt % salt cake can be satisfactorily immobilized by the Exothermic process. Some white smoke and oxides of nitrogen evolve throughout the reaction; the former consists of particulate SiO_2 and Na_2O. Results of laboratory-scale tests of the Exothermic process with synthetic salt cake are described in Ref. (*20*).

Raw material costs for the Exothermic process are higher than those for the Endothermic process. The Exothermic process is also not as

Made from Other Hanford Solid Wastes

Melt Conditions		*Glass Product*	
Temperature	*Time*	*Density*	*Leach Rate in Water (25°C)*
°C	*hr*	*g/cm³*	*g/cm²-day*
1350	0.5	2.5	2×10^{-6} [b]
1350	0.5	2.5	4.5×10^{-9} [d]
1200	0.25	–	6.0×10^{-6} [e]
1050–1200	1.0	–	1.0×10^{-5} [f]

[c] Contained ~ 0.5 mg Pu per gram.
[d] Leach rate (based on Pu) after 98-day leaching.
[e] Leach rate (based on Cs) after 96-hr leaching.
[f] Leach rate (based on Sr) after 96-hr leaching.

amenable to continuous operation as the Endothermic process. A further disadvantage of the Exothermic process is the iron metal which is produced as a by-product. For these reasons the Exothermic process is presently regarded only as a back-up to the Endothermic process and no further laboratory testing of this process is presently scheduled.

Reduction of Salt Cake Volume. Hanford salt cake contains at most only a few milligrams of long-lived ($t_{1/2} > 25y$) radionuclides per metric ton. Any decision to remove the salt cake from the present tanks, immobilize it, and store it elsewhere either onsite or offsite provides strong incentive to concentrate the radionuclides and thereby reduce significantly the volume of inert material which must be immobilized and/or stored.

There are two basically different approaches to achieving a major reduction of the volume of salt cake: (1) removal of sodium, and (2) removal of radionuclides. Exploration of these two alternatives has been very limited so far; such work as has been done is reviewed in the following sections.

REMOVAL OF SODIUM. Workers at the Battelle Pacific Northwest Laboratories (PNL) have recently published (*21*) results of a literature search and preliminary experiments performed to determine the feasibility of reducing salt cake volumes by the removal of sodium and purifying the sodium as metal for reuse or, at least, restricted storage. Methods considered by these scientists for removal of sodium from water solutions of salt cake included:

1. Precipitation of sodium oxalate.
2. Electrolytic reduction after conversion to chloride media.
3. Evaporation, denitration, and chemical reduction of Na_2O with carbon.

Based on their limited work, the PNL investigators conclude that the most favorable sodium removal process would include the following steps:

- Evaporation and denitration.
- Carbon reduction of Na_2O to sodium metal.
- Distillation of sodium metal.
- Purification of sodium metal by filtration and distillation.

They note (*21*), "Considerable development work remains before an integrated process could be demonstrated. Questions include materials of construction to withstand the severe corrosion during the calcination step and the reduction step. Methods of trapping the reduced sodium metal need to be devised. It was found that the nitrate level would have to be reduced to an undetermined low level (probably $< 10\%$) for a safe carbon reduction reaction. Thermal decomposition may have to be supplemented by chemical reduction to achieve this level. Excessive volatility of sodium and cesium compounds was observed during calcination." To this we add that no calculation of the costs of installing and operating such a sodium removal process has been made.

REMOVAL OF RADIONUCLIDES. The obvious converse to removing sodium from the salt cake is to remove the small mass of radionuclides from the large volume of associated inert material. One possible way (Figure 5) of accomplishing the latter objective is to dissolve the salt cake in water and then to use appropriate ion exchange, scavenging, and/or other aqueous separation procedures to reduce the concentration of all long-lived ($t_{1/2} > 25y$) radionuclides to or below those permissible in water in controlled zones and, hopefully, to or below those permissible in water in uncontrolled areas. Subsequently, the small volume of solids and solutions containing the concentrated radionuclides could be immobilized by conversion to glass or, possibly, by the Aqueous Silicate process. The large volume of decontaminated solution could then be stored as a very low-level waste, or, ideally, as a nonradioactive chemical waste.

Some very preliminary tests of this approach have been made with actual salt cake from one tank. In one test 130 g of salt cake were leached 60 min at 25°C to yield 192 milliliters of 7.4*M* Na solution and 2.9 ml of undissolved residue. Approximately 85% of the ^{137}Cs in this salt cake reported to the solution phase, but only 10–20% of the ^{90}Sr and plutonium was water-soluble. The concentration of ^{90}Sr in the salt cake solution was reduced from 31μCi/l. to 0.29 μCi/l. by adding H_3PO_4 and inert $Sr(NO_2)_3$ to precipitate $Sr_3(PO_4)_2$. The resulting liquor containing 5200 μCi/l. ^{137}Cs was passed through three fresh 10-ml beds of Duolite ARC 359 resin, a phenolic-based cation exchanger made by the Diamond Shamrock Co.

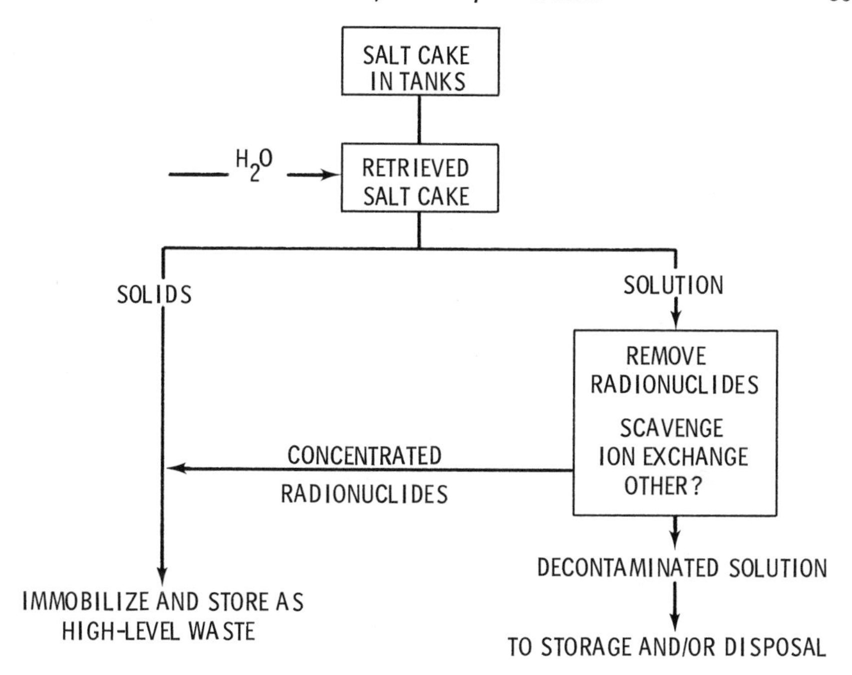

Figure 5. Conceptual salt cake volume reduction scheme

The effluent from the final resin bed contained 0.45 μCi/l. ^{137}Cs and 0.19 μCi/l. ^{90}Sr. These values are well below the presently accepted maximum permissible concentrations of 1 μCi/l. for insoluble ^{137}Cs and ^{90}Sr in water in a controlled zone. Behavior of plutonium in these separation schemes was not established.

It is thought that the scheme indicated in Figure 5 might provide as much as a 10-fold reduction in the amount of high-level waste to be immobilized and stored either on- or offsite. Needless to say, however, neither the technical nor economic feasibility of such a volume reduction scheme has yet been determined.

Acknowledgments

In preparing this review paper we benefited greatly from stimulating discussions with many of our ARHCO colleagues. The advice and counsel of J. S. Buckingham was especially helpful. To him and to all those who gave so generously of their time and ideas we say a heartfelt "Thank you."

Special thanks go to D. E. Larson and P. W. Smith for permitting us to draw so heavily on their detailed accounts of the present Hanford Waste Management Program.

As usual, we are indebted to Eleanore Earhart for her excellent secretarial and editorial help.

Literature Cited

1. Beard, S. J., Kofoed, R. J., Shefcik, J. J., Smith, P. W., "Waste Management Program—Chemical Processing Department," **HW-81481,** General Electric Co.. Richland, Wash., Mar. 1964.
2. Tomlinson, R. E., "The Hanford Program for Management of High-Level Waste," *Chem. Eng. Progr., 60:* Symp. Ser. No. **53,** 2–19 (1964).
3. Lennemann, W. L., "Management of Radioactive Aqueous Wastes from the United States Atomic Energy Commission's Fuel Reprocessing Operations, Experience and Planning," in *Proc. Symp. Manage. Radioactive Wastes from Fuel Reproc.,* Paris, Nov. 27–Dec. 1, 1972, USAEC Report **CONF-721107,** March 1973.
4. Smith, P. W., "High Level Waste Management Program," *in* "Management of High Level Radioactive Wastes at the Hanford Site," Undocumented Report, Atlantic Richfield Hanford Co., Richland, Wash., September 1972.
5. Larson, D. E., "Radioactive Waste Management Program," **ARH-2185,** Atlantic Richfield Hanford Co., Jan. 1971.
6. "Radioactive Waste Management, A Bibliography of Publicly Available Literature Pertaining to the USAEC's Hanford, Washington Production Site," USAEC Report **TID-3340,** Aug. 1973.
7. Barney, G. S., "Fixation of Radioactive Wastes by Hydrothermal Reactions with Clays," **ARH-SA-174,** Atlantic Richfield Hanford Co., Jan. 1974.
8. Krupka, M. C., "Selected Physiochemical Properties of Basaltic Rocks, Liquids, and Glasses," USAEC Report **LA-5540-MS,** Los Alamos Scientific Laboratory, Los Alamos, N.M., Mar. 1974.
9. Leibowitz, L., Williams, C., Chasanov, M. G., "The Viscosity of UO_2-Basalt Melts," *Nucl. Tech.,* **24,** 234 (1974).
10. Saidl, J., Rálkóva, J., "Verfestigung Hochativer Abfälle. 2. Mittlelung: Basalt, Vorteilhaftes. Inkorporieungs—und Fixerungsmediums," *Kernenergie,* **10,** 129 (1967).
11. Kupfer, M. J., Schulz, W. W., "The Endothermic Process—Application to Immobilization of Hanford In-Tank Solidified Waste," USAEC Report **ARH-2800,** Atlantic Richfield Hanford Co., July 1973.
12. Mukerji, J., Kanyal, P. B., "Central Glass and Ceramic Research Institute Report No. 13 and Final Report Part II on Fixation of High Level Atomic Waste in Glass for Ultimate Disposal. Part A: Thermal Decomposition of Fission Products Nitrates and Their Reaction with Glass Batch Additions," Indian Report **BARC-691,** Bhasha Atomic Research Center, Bombay, India, 1973.
13. "Savannah River Laboratory Quarterly Report—Waste Management, July–September 1974," USAEC Report **DPST-74-125-3,** E. I. du Pont de Nemours Company, Savannah River Laboratory, Aiken, S.C., 1974.
14. Vogt, H. G., "Glass" *in* "Encyclopedia of Chemcial Technology," R. E. Kirk and D. F. Othmer, eds., Vol. 7, Interscience Encyclopedia, Inc., N.Y., 1951.
15. Weyl, W. A., "Colored Glasses," Soc. of Glass Tech., Sheffield, England, 1951.
16. Spector, M. L., Suriana, E., Stukenbroeker, G. L., "Thermite Process for the Fixation of High-Level Radioactive Wastes," *Ind. Eng. Chem. Proc. Des. Dev.,* **17,** 117 (1968).
17. Stukenbroeker, G. L., Suriani, E., "Process for the Fixation of High-Level Radioactive Wastes," U. S. Patent **3,451,940,** June 24, 1969.
18. Rudolph, G., Saidl, J., Drobnik, S., Gruber, W., Hild, W., Krause, H., Muller, W., "Lab-Scale R&D Work on Fission-Product Solidification by

Vitrification and Thermite Processes," in *Proc. Symp. Manage. Radioactive Waste from Fuel Reproc.*, Paris, Nov. 27–Dec. 1, 1972, USAEC Report **CONF-721107,** 655, March 1973.

19. Zakharova, K. P., Ivanov, G. M., Kulichenko, V. V., Krylova, N. V., Soroki, Yu. V., Fedorova, M. I., "Use of Heat from Chemical Reactions for the Treatment of Liquid Radioactive Wastes," *Atomnaya Énergiya,* **24,** 475 (1968) through *Soviet Atomic Energy (Engl. Trans.),* **24,** 588 (1968).
20. Kupfer, M. J., Schulz, W. W., "Application of the Hanford Thermite Process to Increase Immobilization of In-Tank Solidified Waste," **ARH-2458,** Atlantic Richfield Hanford Co., Sept. 1972.
21. Burger, L. L., Ryan, J. L., Swanson, J. L., Bray, L. A., "Salt Waste Volume Reduction by Sodium Removal," **BNWL-B-293,** Battelle Pacific Northwest Laboratories, Sept. 1973.

RECEIVED November 27, 1974.

5

The High-Level Radioactive Waste Management Program at Nuclear Fuel Services

JAMES P. DUCKWORTH

Nuclear Fuel Services, Inc., PO Box 124, West Valley, N.Y.

The generation and management of high-level radioactive wastes from both Purex and Thorex type processing have been carried out since 1966. The Purex type waste represents combined high-level waste and intermediate-level waste neutralized and stored in multiconfined mild steel tanks. The Thorex waste which also contains the thorium is stored as an acid solution in stainless steel storage vaults. Future high-level waste management systems include multi-tank complex for five-year cooling and storage as an acidic solution. The location, structure, and multiple containment provides protection against possible releases to the environment. Sketches of the storage systems and tables of waste compositions are included.

The management of high-level radioactive waste at the West Valley, N. Y. Plant of Nuclear Fuel Services, Inc. (NFS) has been another successful first in the private sector of the nuclear power industry. The NFS reprocessing plant is the world's first commercial operation. Construction started in May 1963 and was completed in February 1966. Hot operation of the plant started in April 1966, and since then about 600 tonnes of power reactor fuel exposed to a maximum of 25,000 MWD/MTU and containing about 2000 kg of plutonium have been recovered. This fuel originated in nine different reactors.

The high-level waste treatment system represents the combination of all the experience gained at the various government-operated reprocessing facilities during the previous 20 years. In addition the design incorporated input from the State of New York which had to obtain the necessary statutory authority to enter another first-of-a-kind agreement with NFS and the AEC (now ERDA) for the perpetual care of the high-level radioactive waste. As a result of these inputs, the management of the NFS high-level waste facility has been carried out for more than eight years without a significant operational problem.

The NFS plant was designed to use the standard Purex and Thorex process with two liquid waste streams: (a) low-level liquid waste such as process condensates and cooling water containing only trace quantities of radioactivity which could be discharged directly into the surrounding waterways, and (b) high-level liquids. The NFS reprocessing flow chart (Figure 1) shows that all the process waste streams are concentrated in one of two evaporators. The concentrated bottoms from these evaporators are combined and neutralized for the Purex process and transferred as a slurry to the mild steel high-level waste storage tanks. Therefore, the high level waste at NFS contains both HLW and ILW (intermediate-level waste) such as sodium carbonate–solvent wash solutions. The Thorex wastes are stored as an acidic solution in stainless steel tanks.

A simplified ventilation flow sketch (Figure 2) of the existing NFS high-level liquid waste storage facility shows the layout and relationship of the components. For both types of wastes a spare tank is available to transfer the existing waste in the event the operational tank should leak. The Purex neutralized HLW tank is operated at the boiling temperature by means of an immersion heater since the heat content of the currently stored waste is not sufficient to self boil. This is done to concentrate further the transferred slurries to the control limit of less than $8M$ sodium.

The steam from the tank is condensed in parallel air-cooled condensers, and the concentrate can be recycled for total reflux or diverted to a resin cleanup unit for treatment before being transferred to a holding lagoon for eventual discharge into the local waterways.

The off gas from the condensers consists mainly of air used to circulate the tanks by air-lift circulators; it is passed through a knockout pot, heater, prefilters, and a HEPA (High Efficiency Particulate Absorber) filter before being combined with other process gases for discharge via a 200-ft stack.

The Thorex wastes are not allowed to exceed 140°F; this minimizes corrosion of the stainless steel tanks. A series of three independent cooling coils has proved to be more than adequate for this purpose.

The off gases from the Thorex waste pass through a caustic scrubber to remove the oxides of nitrogen prior to being combined with the Purex waste off-gases. This step is necessary since ammonia is evolved from the Purex waste and the combination of gaseous ammonia with oxides of nitrogen forms ammonium nitrate which sublimes on the filter media

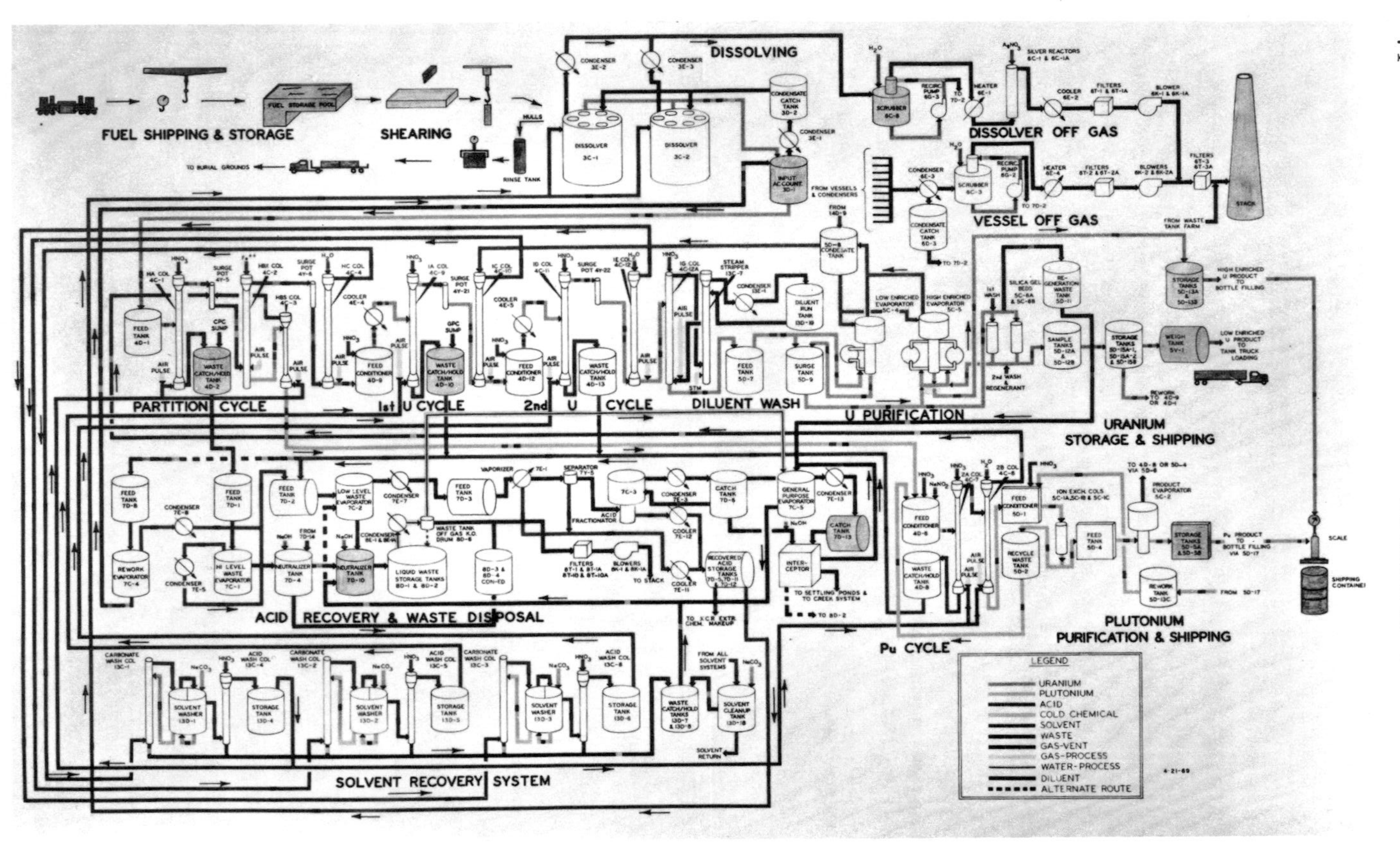

Figure 1. Reprocessing flow chart. Nuclear Fuel Services, Inc., West Valley, N.Y.

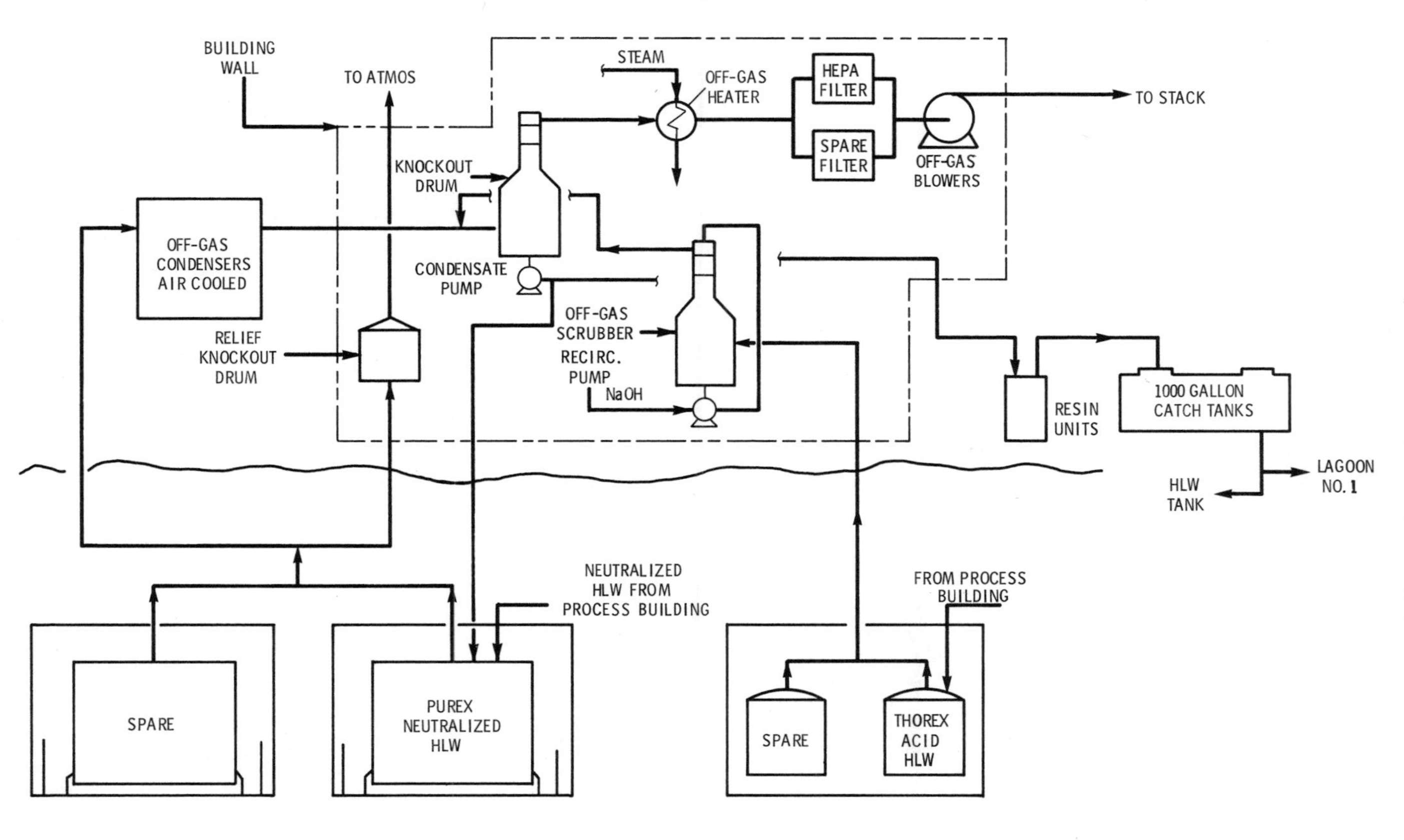

Figure 2. NFS Neutralized high level liquid waste storage facility

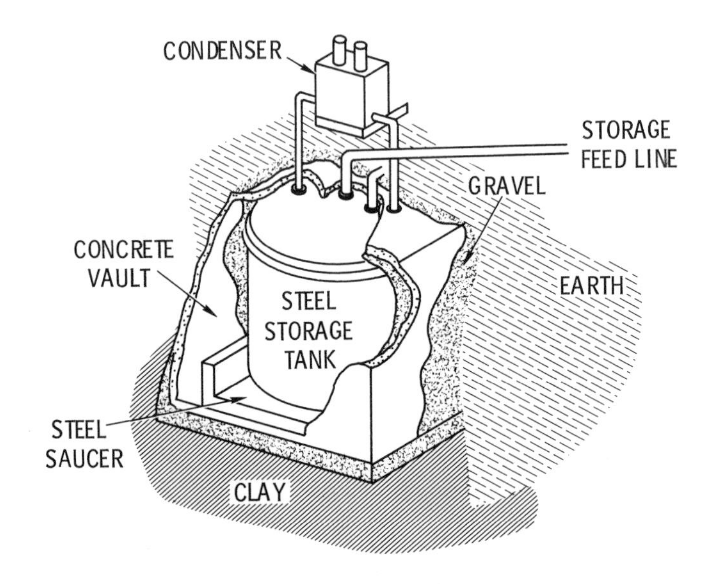

Figure 3. High level waste storage

and plugs the system. This can happen in a matter of hours if the oxides of nitrogen are not completely removed.

Figure 3 is a graphic representation of the multiple phases of confinement that have been incorporated into the NFS HLW storage facilities.

(a) There is a mild steel storage tank with an air-cooled condenser connected directly to it to allow total reflux.

(b) The tank rests on a five-foot-high insulated steel pan which confines any leaks. This arrangement is called a cup and saucer design. The liquid level detectors and liquid transfer pumps which can transfer the leakage back into the tank are connected to the saucer.

(c) The tank and saucer are completely enclosed in a reinforced concrete vault which is covered with more than eight feet of soil. The vault is waterproofed and contains both liquid detectors and removal means similar to the saucer.

(d) The vaults are built below grade in soil that is almost impervious to water flow (horizontal flow rates average 0.005 in. per year). This was one of the major criteria considered when locating the NFS Reprocessing Plant at West Valley, N. Y. Also, the silty till, as it is called, has ion exchange affinities for ^{137}Cs and ^{90}Sr, the two most environmentally significant nuclides in the waste, to the extent of about 90% and 50%, respectively.

(e) Because of the impermeability of the silty till, the HLW vault rests on and is surrounded by four feet of gravel. The gravel is filled with water to a level about equal to the top of the tank by an automatic fill device. This ensures the integrity of the vault by allowing detection of the crack in the concrete by in-leakage. The water level also ensures that the surrounding earth is completely saturated so that if the first four phases were breached, the migration of activity would be restricted further.

(f) The sixth phase is the location of the tanks on a plateau bordered by ravines deeper than the bottom of the vaults. At the migration rates any leakage would require a calculated 40,000 years to reach the closest point of the ravine which would allow almost all of the migrating activity to decay to innocuous levels. However, in addition to this protection, the runoff in these ravines is analyzed routinely as another cross check.

(g) The seventh phase of confinement, which could also be considered the second, is the spare tank present for both types of wastes so that if a leak should occur, a constantly available pump can be installed and the contents transferred to the spare tank. If the spare were used, another spare would be started immediately and funding arrangements for such a contingency are part of the NFS agreement with the New York State Atomic and Space Development Authority, proprietors of the site.

There are many more design and control features to these facilities that aid in their management such as air circulators for mixing the contents, multiple-level and density detectors, temperature measuring devices, radiation alarms, etc. Although we do not take credit for these features as primary phases of confinement of the waste, we do rely on these control systems to ensure the primary barriers or phases are functional.

Table I is a typical daily operating report for the NFS Waste Tank Farm. It shows that temperatures, densities, liquid levels, air flow rates, etc. are observed and recorded at least three times a day. Any equipment malfunction or response to the continuously monitored alarms is reported at least once a day on a supervisor's summary report. With the required initialing of both the operator and the supervisor on each shift, at least six trained individuals are made aware of the operating status of the facilities daily. In addition, these daily reports are distributed to no less than four other responsible members of management for information and review.

In the event of a non-standard or accidental occurrence, the situation is reported verbally to management by a detailed level of authority which extends to the Plant Safety Committee. This Committee is composed of the Plant Manager as Chairman, and the managers of Operations, Technical Services, and Health and Safety. The Plant Safety Committee has the responsibility and authority to approve any action to return the off-standard conditions to a safe controllable condition which includes shutting down the reprocessing plant.

Table I. Waste Tank

Circle Item in Operation:

Condenser:	8E-1	8E-1A
Condenser Fans:	8E-1, N, S	8E-1A, N, S

		SHIFT I		
		TIME	0545	
TANKS				
8D-1	8D-2	8D-1	8D-2	Operator
TI-1	TI-7	72	203	
2	8	72	203	
3	9	72	202	
4	10	72	203	
5	11	72	203	
6	12	72	198	
PR-1	PR-4	−1.2	−1.2	
LI-1	LI-5	6	0	
2	6	21	79	
14	15	9.5	0	
FG-1	FG-5	2	2	
2	6	2	2	
3	7	2	2	
4	8	2	2	
WATER INJ.				
LI-9	PG-10	6.5	0	
TANK OFF GAS				
PRC-3	PRD-2	23	1.5	
CONDENSATE				
TG-4	LG-1	82	55.3	
FG-9	PG-3	off	off	
FRC-8	FRC-10	off	off	
CONDENSER		8E-1	8E-1A	
Shutter	AA	0	0	
Positions	BB	C	C	
	CC (N)	0	0	
	(S)	0	0	
	(E)	0	—	
	(W)	—	0	
Air Temp.				
(60° + 5)	(N)	56	55	
	(S)	55	54	
HEAT LOAD				
8D-2 Coil Pres.		59#		
PG-4	PG-5	52	50	
TRAPS (5)		OK		
CON. ED.		8D-4		
TI-13		106		
TI-14		106		
TI-15		103		
PR-9	LI-11	.5	80	
LI-12	LAH-10	36	off	

Operation Data Sheet

Date ____________________

	8K-1	8K-1A
Blower:	8K-1	8K-1A
Filter:	8T-1	8T-1A

SHIFT II			SHIFT III		
TIME	0805		TIME	2145	
8D-1	8D-2	Operator	8D-1	8D-2	Operator
73	203		72	203	
73	203		72	203	
72	202		72	202	
72	203		72	203	
72	203		72	203	
73	198		72	198	
−1.4	−1.2		−1.5	−1.5	Blew Probes
6	0		5.8	0	Mano.
21	79		21	79.5	
9.7	0		10	0	
2	2		2	2	
2	2		2	2	
2	2		2	2	
2	2		2	2	
6.5	0		6.5	0	
−22	2.7		−25	2.5	
83	55		87	55	
off	off		0	0	
off	off		7	0	
8E-1	8E-1A		8E-1	8E-1A	
0	0		0	0	
C	C		C	C	
0	0		0	0	
0	0		0	0	
0	—		0	—	
—	0		—	0	
56	54		60	58	
55	54		59	58	
59# Air			60# Air		
53	51		53	51	
OK			OK		
8D-4			8D-4		
106			106		
106			106		
104			103		
+.2	80		+.2	80	
36	off		36	off	

Table I.

LI-13	FAL-18	30	off
8G-3	PG-7	on	7.6
8RAH-16	PG-8	off	46
Cooling Coils			
Bottom		H_2O	
Side Inner		Air	
Side Outer		Air	
Shift Supervisor		Initials	

Another phase of the management of HLW at NFS is to ensure that the composition does not exceed specifications. These specifications, shown in Table II along with the actual composition of the Purex wastes as of December 1973, include limits for 10 elements, two ions, excess caustic, heat value, and capacity. The governing specifications for the NFS Purex waste are 8.0*M* sodium plus potassium and the 600,000-gal operational volume of the storage tank.

It is possible to exceed one of the other concentration limits, however. Therefore, besides routine direct sampling and analysis, a detailed system of chemical accountability has been developed at NFS. This was necessary since it is not possible to analyze chemically for all the components shown such as oxalate, C_2O_4.

Briefly, the chemical accountability system is summarized on a

Table II. NFS Stored Neutralized Purex Waste

Component	*Maximum Allowable Composition,*	*December 31, 1973 Composition,*
Na + K	8.0 *M*	7.73 *M*
Fe	0.6 *M*	0.30 *M*
Cr + Ni	0.2 *M*	0.04 *M*
F [a]	0.2 *M*	0.01 *M*
Mo	0.1 *M*	0.01 *M*
Al	0.4 *M*	0.11 *M*
P	0.2 *M*	0.10 *M*
C_2O_4	0.1 *M*	0.09 *M*
Cl	0.1 *M*	0.01 *M*
SO_4	0.1 *M*	0.32 *M*
Excess caustic, %	1 [b]	2.6
Capacity, gallons	600,000	515,000 [c]
Heating value, Btu/hr/gal	200	< 200

[a] Accompanied by at least two moles of Al or Th per mole of F.
[b] Minimum.
[c] Through first quarter 1974.

(Continued)

30	off	30	off
on	6.8	on	6.8
off	46	off	46

H_2O	H_2O
Air	Air
Air	Air
Initials	Initials

monthly report form such as shown in Table III. It includes material balances of warehouse withdrawals vs. inventories, chemical makeups and usage records vs. operating flow sheet rates, input and analysis of auxiliary waste tanks and sump transfers, and calculated corrosion rates vs. chemical analysis and operating time.

The composition of the Thorex waste with the limiting specifications is shown in Table IV. This waste containing the thorium along with the fission products is being maintained at about 120°F using one of three installed coils. Both Purex and Thorex waste systems are being main-

Table III. Waste Surveillance Program

Component	*Moles Beginning of April*[a]	*Moles Added During April*	*Moles at End of April*	*Concentration at End of April, molar*
8D-2 Waste Storage Tank				
Na + K	12,866.0	110.3	12,976.3	6.01
Fe	567.2	7.1	574.3	0.27
Cr + Ni	61.0	0.6	61.6	0.03
F	3.8	0	3.8	<0.01
Mo	10.8	0.1	10.9	0.01
Al	139.6	1.4	141.0	0.07
P	154.9	2.2	157.1	0.07
C_2O_4	90.3	0	90.3	0.04
Cl	7.6	0	7.6	<0.01
SO_4	600.5	4.1	604.6	0.28
OH (Excess)	369.8	0.9	370.7	0.17
8D-4 Waste Storage Tank				
Th	No Transfers This Period		66.81	1.46
Al			16.16	0.35
F			4.7	0.10
HNO			46.79	1.03
P			2.08	0.04
Cl			0	0

[a] 1973.

Table IV. High Level Liquid Waste Storage Thorium Bearing Wastes

Component	*Maximum Allowable Storage Parameter*	*Parameter December 31, 1973*
Th, *M*	2.3	1.46
Al, *M*	0.5	0.35
F, *M*	0.2	0.10
HNO_3, *M*	2.0	1.03
P, *M*	0.2	0.04
Cl, *M*	0.05	0.00
Volume, gallons	13,500	12,047
Temperature, °F	140	100–120

tained under surveillance conditions until ultimate disposal systems are determined.

The expansion program for NFS includes plans to store any future Purex waste as an acid solution for five years before solidifying as required by the 1971 changes in federal regulations. Figure 4 shows a

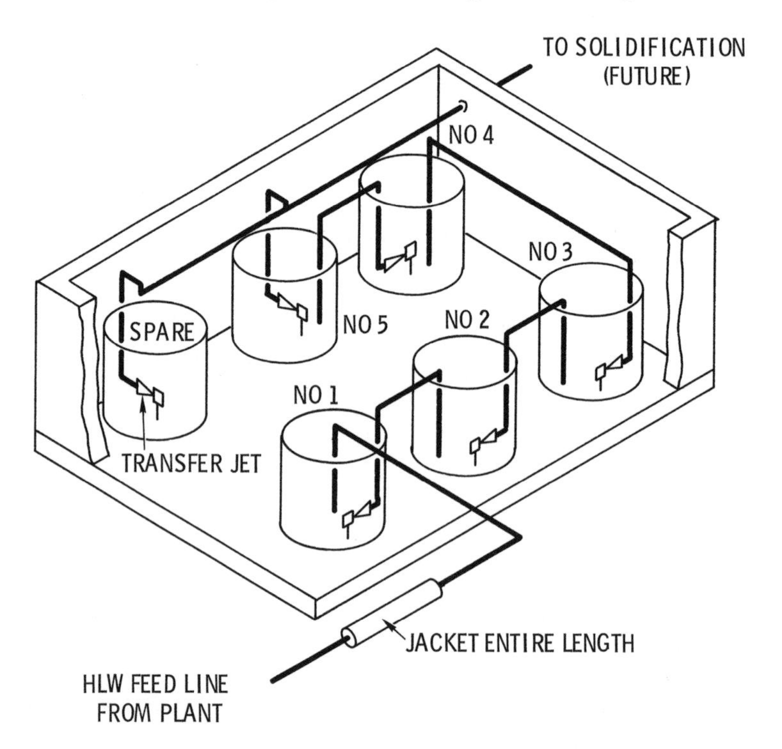

Figure 4. Normal transfer routings for NFS acid HLW storage tanks

simple layout of the proposed NFS HLW acid storage concept. It includes six tanks: a tank for the annual waste for five years of operation and a spare. Because of the decrease in heat content of the decaying wastes and for economic reasons, the heat removal capability of the tank to keep the solution cooled below 140°F will vary from 22 million Btu/hr for Tank 1 to 5½ million Btu/hr for Tank 5. Operationally the tanks will be transferred forward annually with the contents of Tank 5 sent to an HLW solidification facility where it will be solidified, encapsulated, and stored for another five years prior to transfer to a federal repository.

The planned composition for the acidic HLW solutions includes 7.0*M* nitric acid with only trace quantities of sodium and sulfate. The most critical specification for liquid storage will be volume; however, sodium and sulfate content will be the most limiting specification for solidification. The new acidic HLW facility will be designed to meet all the natural phenomena and EPA criteria as well as to include the same degree of confinement conservatism as the existing HLW systems.

This new facility will have two phases of protection added to the overall NFS waste management program: (1) the total quantity of liquid stored will be limited to five years of production, and (2) after five years the liquid will be solidified to a low soluble solid and transferred to a federal repository within 10 years of its production in the NFS plant.

RECEIVED November 27, 1974.

6

High-Level Waste Management Research and Development Program at Oak Ridge National Laboratory

J. O. BLOMEKE and W. D. BOND

Oak Ridge National Laboratory, Oak Ridge, Tenn. 37830

Projections of wastes to be generated through the year 2000 portend a problem of impressive size and complexity but one which can be handled within the framework of current and planned investigative programs. Investigations of the technical feasibility of removing actinide elements from wastes to render the residuals more manageable in terms of hazards and storage requirements indicate that they can be removed from wastes by the minimally desired factors of 10^2 to 10^4; however, demonstrations and engineering assessments of chemical flowsheets have yet to be made. Natural salt formations are believed to offer the best prospects for disposal of high-level wastes; other promising geological formations are also being evaluated for their suitability for use in the disposal of wastes.

Since the mid-1950s, Oak Ridge National Laboratory has been engaged in developing and evaluating waste management methods for the expanding nuclear power industry. Much of this work was a normal outgrowth of our earlier involvement with fuel reprocessing studies since it is in reprocessing that many of the most formidable waste management problems arise. Until recently, the principal developmental approach to the management of high-level wastes was to convert them to stable solid forms suitable for permanent disposal in bedded salt formations. However, within the past year or so, the Energy Research and Development Administration (formerly the U.S. Atomic Energy Commission) has broadened its program in this area to include the evaluation of alternative approaches. Our participation in this new work consists principally of studies to determine the feasibility of separating the actinide elements in the wastes from the fission products for the purpose of reducing the long-term risk in disposing of residual fission products, and evaluations of geologic formations other than bedded salt for their stability as disposal media.

Waste Projections

Although all forecasts of this nature are inexact and even controversial, they nevertheless can serve as the best available guide to future conditions and thus have become a useful tool for planning and evaluating purposes. Recently, we completed projections of all the wastes that would arise from the nuclear fuel cycle over the remainder of this century (*1*); however, only our estimate of high-level wastes is presented here.

The projections are based on a recent forecast (Case B) by the Energy Research and Development Administration (ERDA) of nuclear power growth in the United States (*2*) and on fuel mass-flow data developed for light water reactors fueled with uranium (LWR-U) or mixed uranium and plutonium oxide (LWR-Pu), a high temperature gas-cooled reactor (HTGR), and two liquid-metal-cooled fast breeder reactors (LMFBRs). Nuclear characteristics of the fuels and wastes were calculated using the computer code ORIGEN (*3*).

The projected installed nuclear electric capacity through the year 2000 is shown in Figure 1. The total installed nuclear capacity is expected to rise from 5.9 GW at the end of 1970 to 102 GW in 1980 and to 1200 GW in the year 2000. This represents 1.7%, 15%, and 54%, respectively, of the nation's total electric generating capacity in those years. In this projection, the greatest proportion of the nuclear capacity is contributed by LWRs. Plutonium, in the form of mixed plutonium and uranium oxide fuels, is recycled in LWRs beginning in 1977, and these mixed oxide fuels represent about 10% of all LWR fuels through the year 2000. HTGRs come on-stream in 1980. The LMFBRs go on-stream in 1987, and constitute almost 200,000 MW of installed capacity by the end of the century.

Following their discharge from the reactors, the irradiated (spent) fuels are shipped to reprocessing plants where the uranium and plutonium are recovered for use in fabricating new fuels. The fission products, residual uranium and plutonium (i.e., losses), and heavy elements that are formed by transmutation in the reactors appear as radioactive constituents in the high-level liquid wastes from the primary chemical separation. We estimate that the metric tons of spent fuels to be reprocessed annually will increase from 1700 in 1980 to 22,000 in the year 2000. Generation of concentrated high-level liquid waste will increase from 170,000 gal/year in 1980 to 2.2 million gal/year in 2000; and, if we assume that this material is stored as a liquid for five years before being solidified, about nine million gal of liquid waste will be stored and an additional two million gal of solidified waste will have been accumulated by the year 2000 as a function of time following that date. We have assumed that the actinide elements in the waste consist of 0.5% of the uranium and plutonium originally in the spent fuels plus all other heavy elements formed by transmutation of the uranium and plutonium in the reactors. The inhalation and ingestion

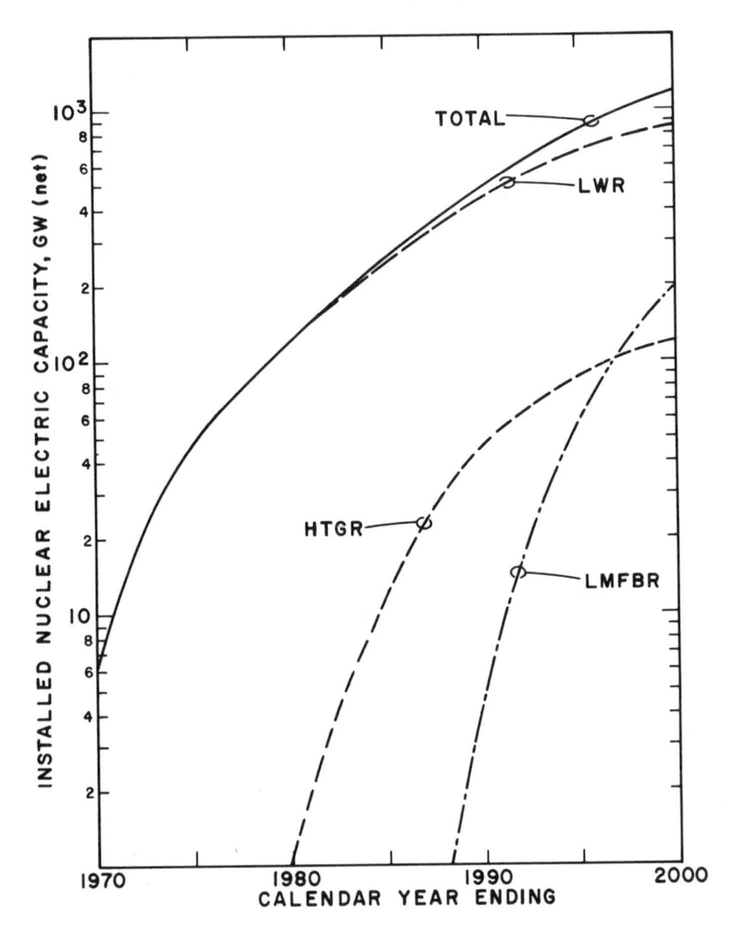

Figure 1. Projected installed nuclear capacity in the United States

toxicities are defined as the cubic meters of air or water that would be required to dilute the radioactive constituents to levels specified in the Radiation Concentration Guides (RCG) (*see* Column 2 of Tables I and II, Code of Federal Regulations, Title 10, Part 20) as the maximum allowable for unrestricted use. The fission products predominate in terms of Curies, heat generation, and ingestion toxicity for the first several centuries; however, after a decay period of 1000 years, the actinides control in every instance. On this basis of the inhalation toxicity, the actinides predominate even initially. This fact has led to the question, why not remove the actinides from the fission products in the wastes and thus reduce the requirement for containment from a million or more years to only 1000 years or less? The separated actinides could be recycled to

Table I. Properties of High-Level Wastes to be Accumulated by the Year 2000

	Accumulated by Year 2000	*Time Elapsed Following Year 2000, Years*			
		1000	*10,000*	*100,000*	*1,000,000*
Radioativity, MCi	158,000	63.1	24.5	4.7	1.6
fission products	155,000	4.7	4.5	3.4	0.7
actinides	3,400	58.4	20.0	1.3	0.9
Thermal power, kW	770,000	1,500	470	36	21
fission products	657,000	5.3	4.9	3.1	0.2
actinides	113,000	1,500	470	33	21
Inhalation toxicity, m^3 air	1.2×10^{22}	3.4×10^{20}	1.4×10^{20}	1.0×10^{19}	4.4×10^{18}
fission products	7.0×10^{20}	1.8×10^{15}	1.8×10^{15}	1.3×10^{15}	2.1×10^{14}
actinides	1.1×10^{22}	3.4×10^{20}	1.4×10^{20}	1.0×10^{19}	4.4×10^{18}
Ingestion toxicity, m^3 water	5.2×10^{16}	1.1×10^{13}	3.3×10^{12}	1.7×10^{12}	6.4×10^{11}
fission products	5.2×10^{16}	1.8×10^{10}	1.8×10^{10}	1.4×10^{10}	2.4×10^{9}
actinides	4.4×10^{14}	1.1×10^{13}	3.3×10^{12}	1.7×10^{12}	6.4×10^{11}

reactors as fresh fuel or as special targets for burnup to shorter-lived fission products.

Waste Partitioning

Figure 2 illustrates the possible merits of such separations by comparing the hazard index (the hazard index is defined as the volume of water that could be contaminated to the RCG-allowable value by one volume of waste or other radioactive material) of the wastes generated by conventional processing of fuel from a typical LWR with that of the wastes resulting from a postulated secondary treatment in which the five principal actinides (uranium through curium in the periodic table) are removed by factors of 10^3–10^5. It is also assumed that 99.9% of the iodine has been separated during reprocessing. For reference, this figure shows the hazard index associated with the mineral pitchblende and with an ore containing 0.2% uranium which is typical of the large deposits that occur on the Colorado Plateau. Note that the hazard index of the waste from conventional processing decreases rapidly over the first 1000 years (because of the decay of ^{90}Sr and ^{137}Cs) but remains as great as or greater than pitchblende for more than a million years. The waste resulting from secondary treatment, however, falls within the range of naturally occurring radioactive materials after only several hundred years, a time span for which we can reliably extrapolate the effects of geologic, climatic, and

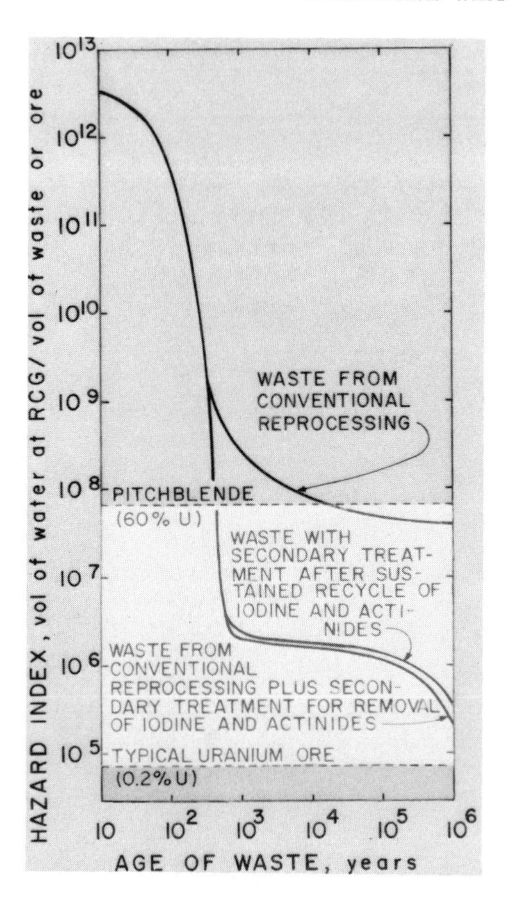

Figure 2. Effect of age and method of treatment on the hazard index of high-level wastes from LWR's

other natural phenomena. It is reasonable to expect that this type of waste could be buried in carefully selected geologic formations with negligible risk to future generations. One way of disposing of the separated actinides and iodine would be to recycle them with fresh fuel to power reactors. Our calculations show that, even after sustained recycling of these materials, the wastes, while having a slightly higher index, would still lie well within the range of naturally occurring radioactive species (*4*).

While Figure 2 was constructed to illustrate the maximum effect on the hazard index that could be achieved using present separations technology, a more realistic goal is to separate each actinide only to a degree which is in proportion to its contribution to the total hazard. Thus, for the hazard index of the wastes to be reduced to less than 5% of that of pitchblende after storage for 1000 years, we must recover 99.99% of the plutonium, 99.9% of the uranium, americium, and curium, and 95% of the neptunium from the spent fuels (Table II) (*5*). In addition, 95% of the thorium and protactinium must be recovered when reprocessing HTGR fuel. Further, reduction of actinides in the wastes would not be warranted unless the very long lived fission products, ^{99}Tc, ^{93}Zr, and ^{135}Cs were also removed along with some additional ^{129}I.

Separations processes currently used to recover actinides for commercial or scientific uses have yielded uranium and plutonium recoveries of about 99.5%, and neptunium, americium, and curium recoveries in the range of 90–99%. It is therefore apparent that a comprehensive development program will be required to reduce this proposed waste management concept to practice. This program must be designed to (a) solve problems that are obvious from previous experience, (b) increase the removal of actinides to the desired levels, (c) determine the most desirable method for integrating the needed process cycles into an overall system, (d) choose chemical processes and reagents that minimize high-level waste treatment problems, and (e) determine the composition of all waste streams that will be generated and ways to recycle these streams.

We have begun our investigations by limiting our considerations to wastes from reprocessing LWR fuels because this is the only type for which experience on large scale reprocessing exists. Concentrations of key constituents in a typical waste as well as required recovery factors of the actinides from that waste are shown in Table III. We have constructed a variety of flowsheets based on previously established separations processes, considering it important that waste streams generated by processes for recovery of the actinides from the wastes be suitable for recycle to previous stages or for combination and management with the residual fission-product waste. To date, only limited laboratory experiments with synthetic solutions of fission products have been performed. Work with tracer levels of actinides will get under way shortly.

Table II. Percent Removals From Spent Fuels Required to Reduce the Hazard Index of Aged Wastes to Approximately 5% of That of Pitchblende

	Type of Spent Fuel			
Actinide	*PWR-U*	*PWR-Pu*	*LMFBR*	*HTGR*
Th	—	—	—	95.5
Pa	—	—	—	95.0
U	99.9	99.9	99.9	99.99
Np	95.0	95.0	70.0	95.0
Pu	99.95	99.99	99.99	99.9
Am & Cm	99.9	99.9	99.9	95.0

Table III. Composition of High-Level Waste from Reprocessing LWR Fuel, and Actinide Recovery Factors Required[a]

Constituent	*Concentration, g/l.*	*Recovery Factor Required*
U	0.93	≥ 5
Np	0.09	≥ 20
Pu	0.009	≥ 50
Am	0.03	≥ 1000
Cm	0.007	≥ 1000
Total fission products	5.60	
Rare earths	2.03	

[a] Basis: 33,000 MWd/ton exposure; 5175 l. waste per ton of fuel; 0.5% loss of U and Pu to waste.

Our present conceptual flowsheets are based on the general approach shown in Figure 3. Uranium, neptunium, and plutonium would first be recovered to the desired degrees using successive tributyl phosphate extraction stages. The problems to be resolved here are concerned with (a) keeping neptunium in the extractable +4 or +6 valence states and (b) minimizing the effect of solvent degradation products and other impurities on plutonium recovery and on later separations.

Next, the waste would be stored for about five years to reduce radiation damage to process reagents, solvents, or ion exchange resins to be used in subsequent separations. The problems associated with this step are that large inventories of waste in liquid form are accumulated and that plutonium grows back into the waste during storage and may again require removal.

Following storage, americium and curium are separated using either the Talspeak (*6*) or the CEC (*7*) process. The Talspeak process is based on solvent extraction with an organic phosphoric acid from a complexed carboxylic acid solution while the CEC process uses cation exchange chromatography. Both processes are capable of separating americium and curium from the lanthanides in the absence of other fission products. The problem, however, is to prepare a suitable feed for Talspeak or CEC by first separating americium, curium, and the rare earths from the other fission products. We are considering two solvent extraction processes for this purpose—one using undiluted tributyl phosphate and the other using di(2-ethylehexyl) phosphoric acid.

As indicated above, much work must be done before even technical feasibility can be established, much less for the probable economic impact to be estimated. After feasibility is established, we envision that another 10–15 years will be required for plant scale process development and demonstration of this waste management approach.

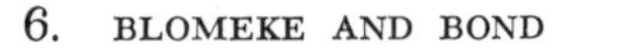

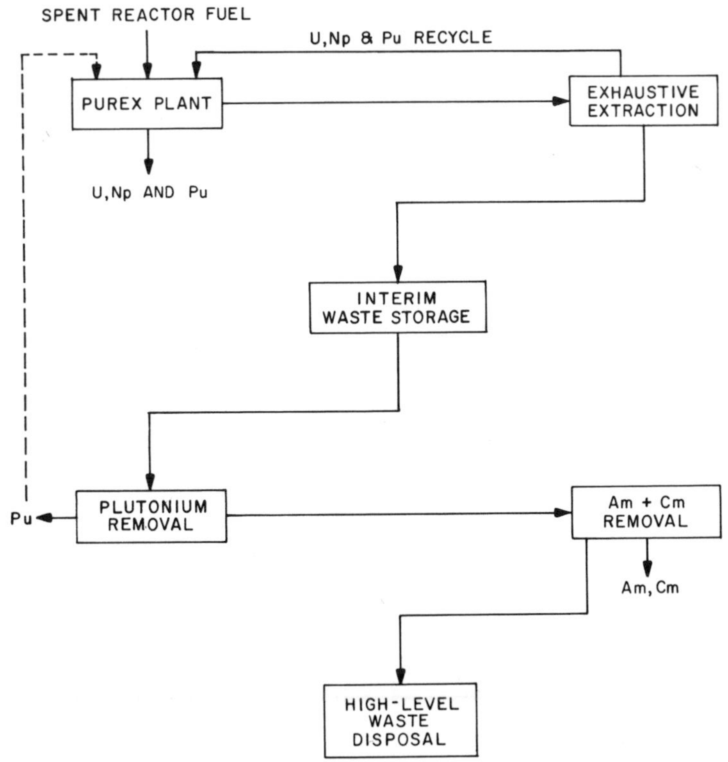

Figure 3. Conceptual processing sequence for the removal of actinides from LWR reactor fuel

Geologic Disposal Evaluations

Following the unsuccessful attempt to establish a demonstrational waste repository near Lyons, Kansas during 1970–71, the AEC (now ERDA, Energy Research and Development Administration) asked us to look elsewhere for a site where a pilot-plant repository might be established to gain final confirmation of the suitability of bedded salt for disposal of high-level waste. With the help of the U.S. Geological Survey and others, an area in the southwestern part of the Permian Basin in southeast New Mexico was chosen for intensive investigation.

We have drilled two 4000-ft-deep exploratory holes at a site about 30 miles east of Carlsbad and 40 miles southwest of Hobbs. Hydrological tests have been made, and geophysical logs of both holes were taken. Cores were recovered for mineralogical studies and physical properties measurements. Additional investigations under way in New Mexico include a local microseismic study, and appraisals of the oil and potash potentials of the area of interest.

Finally, we have established a program designed to clarify the mechanism of salt dissolutioning by ground water in and around boreholes penetrating a salt formation. We are also sponsoring work on methods and materials for use in permanently plugging boreholes and mine shafts. The need to develop and demonstrate highly reliable methods for sealing all man-made penetrations is a requirement for disposal into any type of geologic formation, but it is most acute for salt because of that material's high solubility in water.

Thus far, all of our work has yielded encouraging results, and we feel as confident as ever that bedded salt is completely suitable for the disposal of solidified high-level wastes either with or without actinides. Furthermore, our relationship with both state officials and private citizens in New Mexico has been very good. However, the ERDA has made a recent decision to avoid commitments to any type of geologic disposal in the immediate future; as a consequence, we have been asked to broaden our investigative program to include an evaluation of other promising geologic formations for storage or disposal of not only high-level but of other types of wastes as well. The formations of greatest interest are salt anticlines and piercement domes, shales, granite, and limestone. We are only now reorienting our program and formulating a detailed plan to achieve the desired objectives. Nevertheless, this clearly represents an intention on the part of the Energy Research and Development Administration to develop a number of viable options for the possible future geologic disposal of many of the wastes that will be generated over the next several decades.

Literature Cited

1. Blomeke, J. O., Kee, C. W., Salmon, R., "Projected Shipments of Special Nuclear Material and Wastes by the U.S. Nuclear Power Industry," USAEC Report **ORNL-TM-3965,** Oak Ridge National Laboratory, August 1974.
2. U.S. Atomic Energy Commission, "Nuclear Power Growth 1974–2000," USAEC Report **WASH-1139(74),** February 1974.
3. Bell, M. J., "ORIGEN—The ORNL Isotope Generation and Depletion Code," USAEC Report **ORNL-4628,** Oak Ridge National Laboratory, May 1973.
4. Claiborne, H. C., "Neutron-Induced Transmutation of High-Level Radioactive Waste," USAEC Report **ORNL-TM-3964,** Oak Ridge National Laboratory, December 1972.
5. Claiborne, H. C., "Effect of Actinide Removal on the Long-Term Hazard of High-Level Waste," USAEC Report **ORNL-TM-4724,** January 1975.
6. Weaver, B., Kappelmann, F. A., "Preferential Extraction of Lanthanides over Trivalent Actinides by Monoacidic Organophosphates from Carboxylic Acids and from Mixtures of Carboxylic and Aminopolyacetic Acids," *J. Inorg. Nucl. Chem.* (1968) **30,** 363–72.
7. Wheelwright, E. J., Roberts, F. P., "The Use of Alternating DTPA and NTA Cation-Exchange Flowsheets for the Simultaneous Recovery and Purification of Pm, Am, and Cm," **BNWL-1072,** Battelle Northwest Laboratory, July 1968.

RECEIVED November 27, 1974. Work was sponsored by the Energy Research and Development Administration under contract with the Union Carbide Corp.

7

High-Level Radioactive Waste Management Research and Development Program at Battelle Pacific

JOHN E. MENDEL, JACK L. McELROY, and ALLISON M. PLATT

Battelle Pacific Northwest Laboratories, P. O. Box 999, Richland, Wash. 99352

Solidified waste forms for the immobilization of high-level radioactive wastes from the commercial reprocessing of power reactor fuels, and processes for the reliable production of the waste forms, are being developed at Battelle-Northwest. The development program has started out on a nonradioactive laboratory and pilot plant scale, and will be carried through fully radioactive engineering-scale demonstrations of the processes. Current development is emphasizing silicate glass or glass-ceramic forms, produced in a two-step, calcination-melting process. Depending upon the waste composition, the glasses or glass-ceramics can contain over 25 wt % fission products and 2 wt % actinides, creating unique problems in assuring long-term stability.

The goal of the high-level radioactive waste management research and development program at the Pacific Northwest Laboratory (PNL), operated by Battelle Memorial Institute, is to provide technology to convert highly radioactive liquid waste, generated as a by-product of nuclear power, to a chemically-inert solid form. The primary emphasis is on commercial wastes; i.e., PNL is devoted to perfecting methods for immobilization of the high-level wastes that will result from reprocessing power reactor fuels in industry-operated reprocessing plants. Currently, four such industrial plants are either built, under construction, or still in the planning stages. These are (a) the Nuclear Fuels Services plant at West Valley, N. Y., (b) General Electric's Mid-west Fuels Reprocessing Plant at Morris, Ill. (operation of this plant has been postponed indefinitely), (c) the Allied General Nuclear Services plant at Barnwell, S. C., and (d) a plant to be operated by the Exxon

Nuclear Corporation and tentatively situated in Tennessee. It has been estimated that at least eight industry-operated fuels reprocessing plants will be operational by the year 2000. The high-level wastes from these plants will be at least one order of magnitude more radioactive than the wastes now being handled routinely at AEC (now ERDA) reprocessing facilities.

The processes being developed at PNL convert the commercial high-level wastes to glasses or related ceramic forms. These materials offer the best practicable immobilization of radioisotopes, in a highly concentrated form, available today; the AEC is also sponsoring continuing research on potentially more advanced solidified waste forms which may become available at a later date and offer added increments of safety or processing economy.

Background

Why Glass? Glass or glass-like forms are being emphasized in the AEC's development program for the near-term immobilization of power reactor wastes because they have the best proven durability of man-made materials. The earliest known man-made glass dates from the 23rd Century B.C. (*1*). However, much older glass occurs naturally. The survival of this glass, plus the studies that have been made on weathered glass surfaces, lend confidence to our ability to predict the future integrity of glass. For instance, by microscopic examination it has been possible to correlate the age of man-made glass and naturally occurring obsidian with the depth of surface alteration. This has been done for glass over 1500 yr old (*2*) and for obsidian surfaces up to 200,000 years old (*3*).

Glass is a very congenial material. Although very few of the oxides, SiO_2, P_2O_5, B_2O_3, GeO_2, As_2O_3, are glass formers by themselves, much of the periodic table can be incorporated into oxide glass as either network formers or modifiers. This broad acceptance benefits the immobilization of high-level wastes, which are complex mixtures of about 35 fission product elements and 18 actinide and daughter elements.

PNL Involvement in Treatment of High-Level Waste From Nuclear Power Generation. Experimental programs for the development of immobilization procedures for the high-level waste from power reactor fuel reprocessing began in several AEC laboratories about 15 years ago, when nuclear power was in its infancy. The engineering-scale solidification process demonstration work has been centered at PNL since about 1965. Prior to that time, the AEC's experimental program had led to the development of three candidate processes for conversion of HLW wastes to calcines, microcrystalline solids, or glasses. These three processes were compared on a fully radioactive engineering scale

Table I. Summary of the WSEP Program Conducted at PNL Between 1966 and 1970

Product Form	*Process*	*Number of Canisters*	*Total Volume of Solid Produced*	*Total Megacuries Encapsulated*
Calcine	Pot Calcine (Developed at ORNL)	8	590	6.6
Phosphate	Phosphate glass (Developed at BNL)	11	660	19.1
Phosphate ceramic	Spray-calciner melter (Developed at Hanford)	12	670	20.9
Borosilicate glass	Rising level glass and in-pot melting (Developed at ORNL and Hanford)	3	120	5.6

in a series of parallel demonstration runs in the Waste Solidification Engineering Prototypes (WSEP) program at PNL. To date, the WSEP program is the most comprehensive demonstration of the conversion of commercial HLW to solid forms. The timing of this program was such that representative commercial HLW had not been produced. The waste feed stream for the WSEP demonstrations had to be manufactured using AEC wastes plus large concentrations of additional radioisotopes (mainly ^{144}Ce) to obtain the heat generation rate of commercial wastes.

Thirty-three canisters of solidified HLW were produced in the WSEP Program (*3*). Over 52,000,000 Ci were incorporated in 2000 l. of solids contained in canisters 6–12 in. in diameter and 8 ft long (Table I). The solid forms produced were (a) oxide calcine made by the pot calcination process developed at ORNL, (b) phosphate glass made by a Brookhaven National Laboratory-developed process, (c) microcrystalline product made by the spray solidification process developed at Hanford, and (d) borosilicate glass made toward the end of the demonstration program in the pot calcination and spray solidification process equipment. Evaluation of processing operability and characteristics of the solidified waste solids led to the conclusion that processes producing borosilicate glass were preferred over the other processes. A new development program, the Waste Fixation Program (WFP), is now underway at PNL to optimize silicate glass fixation since only a modest effort toward development of borosilicate glass was included in the WSEP program, and the WSEP borosilicate glass processes utilized a low processing temperature (900°–950°C) which produced glass of less than optimum quality.

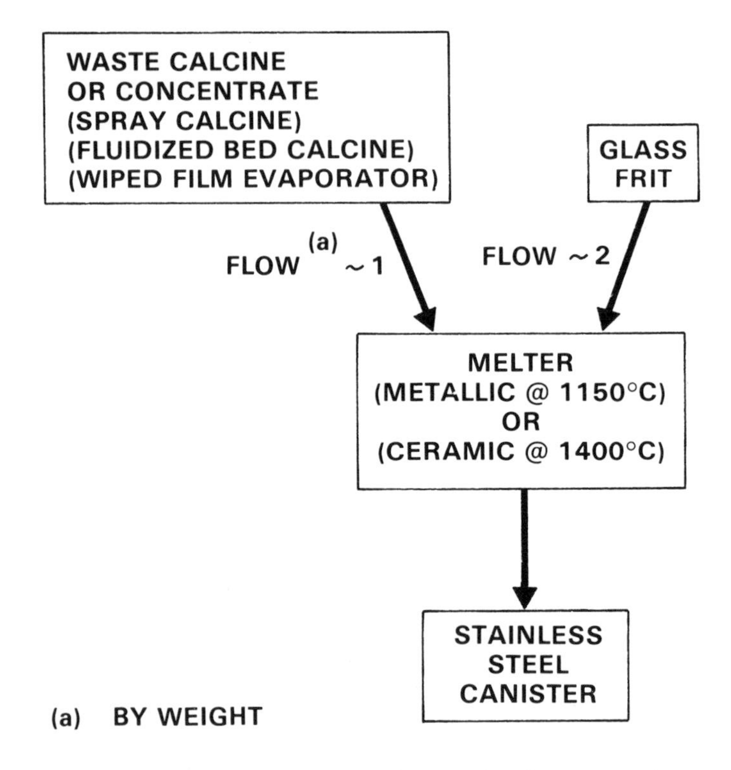

Figure 1. Basic glassmaking flowsheet utilized in waste fixation program

PNL Waste Fixation Program

The WFP Solidification Process. The WFP program emphasizes the development of the technology needed to incorporate commercial HLW into silicate glass or glass-like materials that will meet strict specifications concerning physical stability and leach resistance. The basic process demonstrated in the WFP is simple; it consists of drying the waste, melting it after the addition of glass-forming chemicals, and casting the molten glass into stainless steel canisters. Several options for each processing step are being evaluated in the development program, as shown in Figure 1.

Glass compositions for the WFP processes are currently being studied, and operation of the processing steps is being perfected in the nonradioactive pilot plant.

Waste Compositions Selected for Study in the WFP. The major source of high-level wastes is the raffinate or aqueous waste stream from the first cycle of the Purex solvent extraction process. The stream is a nitric acid solution containing over 99.9% of the nonvolatile fission products present in the irradiated fuel elements, traces of iodine, iron in some instances, traces of uranium and plutonium, and all the other actinide elements produced in nuclear power reactors. It will also contain iron, chromium, and nickel from the corrosion of stainless steel, and a small amount of phosphate from the phosphate-containing extractant used in the solvent extraction process. However, the high-level waste stream coming from the reprocessing plant is not made up of the first-cycle raffinate exclusively. It can also contain complements from subsequent solvent extraction cycles, and various scrubber and decontamination wastes, which are not so well characterized as to quantity and composition. These may contribute relatively large amounts of sodium.

To demonstrate that the various potential high-level waste compositions are all amenable to immobilization, the WFP program uses two standard waste compositions. These are a clean waste, designated PW-4b (which is essentially first-cycle raffinate), and a dirty waste (designated PW-6), which contains large amounts of sodium and iron (*see* Table II). It should be noted that there are other possible constituents of high-level wastes, such as fluoride, mercury, and soluble poisons which are not included in the initial demonstration program.

During waste treatment constituents may be added which will affect subsequent fixation processes, e.g., the inert seed particles used in fluidized bed calcination. These particles, which may at times comprise as much as two thirds (by weight) of the waste calcine, must become part of the waste glass. Because of their potentially large perturbation on glass-making, studies involving calcine with seed particles are included in the WFP program.

Silicate Melt Formulations

The silicate melt formulations being used for the fixation of radioactive wastes are quite different from industrial glass formulations. Thus, although the principles developed through the millennia that man has produced glass are used as guidelines, the glass formulations used for waste fixation are being developed empirically.

Examples of the candidate glass formulations are shown in Table III. These are developed for the reprocessing of PW-4b, PW-6, and calcine with seed particles in the WFP equipment. As shown in Figure 1, two melter concepts are being developed in the WFP program, one operating at 1150°C and one operating at ca. 1400°C. The latter permits use of a higher-melting higher-silica content glass formulation which more closely approximates industrial glass compositions. The properties of these glass formulations are being determined in detail. Although continued testing undoubtedly will reveal desirable modifications, present indications are

Table II. Waste Compositions Selected for Waste Fixation Program

Constituents[a]	*Concentration,* M, *at 378 l./MTU*	
	PW-4b	*PW-6*
Process inerts		
H	1.0	3.0
Na	—	2.0
Fe	0.05	0.5
Cr	0.012	0.045
Ni	0.005	0.020
PO_4	0.025	0.025
Fission products		
Mo	0.095	0.095
Tc (Mo)	0.022	0.022
Sr	0.027	0.027
Ba	0.027	0.027
Cs (K)	0.054	0.054
Rb (K)	0.010	0.010
Y + RE (RE mix)	0.208	0.208
Zr	0.106	0.106
Ru (Fe)	0.059	0.059
Rh (Co)	0.010	0.010
Pd (Ni)	0.032	0.032
Te	0.012	0.012
Ag	0.002	0.002
Cd	0.002	0.002
Actinides		
Np (U)	0.0085	0.0085
U + Pu (U)	0.011	0.054
Am Cm	0.0022	0.0022

[a] Chemical stand-ins used are shown in parentheses.

that glass formulations of the types in Table III can serve as satisfactory fixation media for high-level wastes.

Properties of the Silicate Melt Formulations

The glass formulations are optimized to provide low leachability and high integrity when solidified, and yet permit trouble-free processing as a melt in remotely operated equipment. The most important properties are:

Processing

- viscosity
- homogeneity
- corrosivity
- volatility

Storage and Disposal

- chemical inertness, leachability, and canister corrosion
- homogeneity
- physical stability: long-term thermal effects, thermal shock, physical shock, and radiation effects.

Properties Required During Processing. The processing equipment places specific requirements upon the properties of the molten waste product, e.g., in the present WFP equipment the molten glass is drained out of the melter periodically through a 1/2-in. drain spout; a melt viscosity of $\leq$ 200 poises is required to do the melt discharging in a controlled and timely manner.

Several of the many constituents of the waste glass are not completely dissolved in the current flowsheets using a metallic melter and a 1150°C-melt temperature. Microcrystallites of the fission products CeO_2 and RuO_3 can settle gradually in the melter, resulting in a higher viscosity melt at the drain point, which makes draining of the melter more difficult. For this and other reasons (melter throughput is increased), the WFP metallic melter is being operated with an agitator.

Corrosion of the glass-making melters must be maintained at an absolute minimum to increase the lifespan of the melter. Laboratory-measured corrosion rates indicate that melter lifetimes of several years can be achieved with high chrome oxide or zircon refractories; metallic melters may have lifetimes of several months if alloys such as Inconel 690 are used. These conclusions have been reached on the basis of extrapolation of laboratory tests. Long-term tests, particularly with waste glasses in engineering-scale continuous melters, have not yet been made.

Table III. Typical HLW Glass Compositions

	Low Temperature Glass (for Metallic Melter operating @ 1150°C)	*High Temperature Glass (for Refractory operating @ 1400°C)*
SiO_2, wt %	27–34	50–53
B_2O_3	10–12	—
Na_2O, K_2O	8–9	5–6
ZnO	19–22	11–13
Fe_2O_3	2–6	7–12
CaO, MgO, SrO, BaO	5–6	—
FP oxides	12–23	11–14
Miscellaneous (actinide oxides, NiO, Cr_2O_3, P_2O_5, etc.)	2–4	2–4

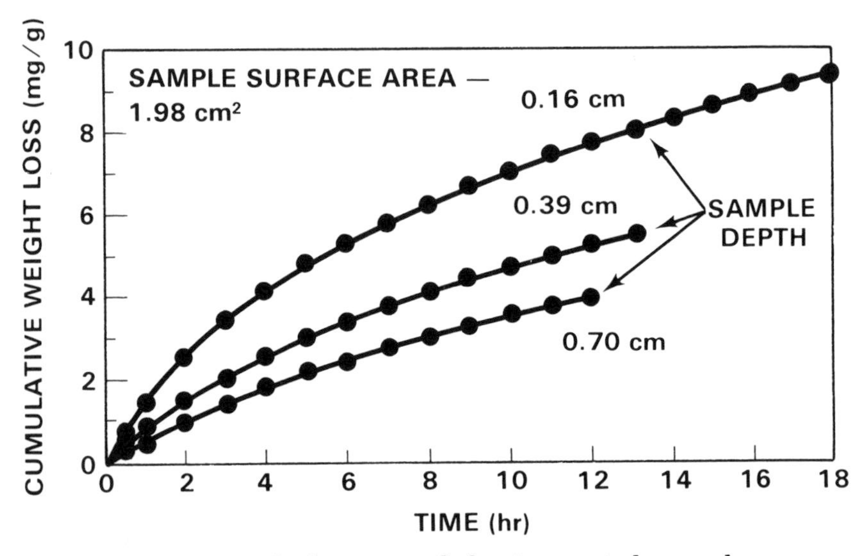

Figure 2. Volatilization weight loss from typical waste glass

Volatility from the high temperature melts can complicate the off-gas treatment; the policy is to recycle volatiles back to the process. For some elements, such as cesium and tellurium, this is a simple procedure. However, volatilized ruthenium, for example, tends to plate out on stainless steel surfaces as difficult-to-remove RuO_3; therefore, the goal is to minimize volatility. Rates of volatilization of fission products from a representative WFP waste glass are shown in Figure 2. Cesium, tellurium, ruthenium and molybdenum account for 95 wt % of the volatility at 1200°C. The relative quantity volatilized is dependent on melt geometry. As the melt gets deeper, as it will in the engineering-scale equipment, the relative quantity volatilized decreases. An additional factor which will help control volatility in the engineering-scale equipment will be the "cold cap" which is to be used in the high temperature melter. This melter will be heated by passing an electric current through the molten melt. The waste calcine and glass frit will be added at the top of the melter where they will float as a "cold cap" to trap the volatiles from the hotter melt below. This is a standard operating practice in the glass-making industry.

Properties Required During Storage and Disposal. The purpose of waste fixation processing is to provide the safest, most trouble-free containment. The radioactive waste glasses will be cast in stainless steel canisters. The solidified waste "package" (i.e., the stainless steel canister and its contents) is considered the integral product of the waste solidification process. Since the stainless steel canister is an integral part of the product, its integrity should not be affected by interaction with the glass, either during the casting operation (when the glass in the canister is temporarily molten), or during storage. Measurements indicate that the corrosion of stainless steel in waste borosilicate glasses is less than one mil/month at 900°C and rapidly becomes immeasurable as the temperature decreases below 900°C. External factors govern the lifetime of the stainless steel canister.

If anything should happen to the stainless steel canisters, the properties of the waste glass become the ultimate control on the rate of release of radioactivity to the environment. The most probable mechanism for release to the environment is dissolution in water. Thus, determination of water-leaching behavior is one of the principal factors considered in the evaluation of candidate solidified waste forms. The leaching of constituents from glass is a complex phenomenon. One obvious approach is to make comparisons with glass of known stability; this is done in Figure 3 where the leach rates of candidate waste glasses are compared with a standard soda lime glass and with borosilicate glass. Waste glasses can be made in metallic melters which compare quite well with soda lime glasses. If a high temperature refractory melter is used, the leach rates of waste glass can approximate those of borosilicate. The leach rates shown in Figure 3 are for nondevitrified glass. As discussed later, devitrification of at least portions of the waste glass probably cannot be avoided. Indications are that devitrification increases the leachability of borosilicate waste glass, but not by more than a factor of ten.

The quantity of radioactivity released if the glass product contacts water is a function not only of leachability, but also of surface area exposed. Thus, the glass should be maintained in large pieces to minimize the surface-to-volume ratio. Since glass is brittle, management

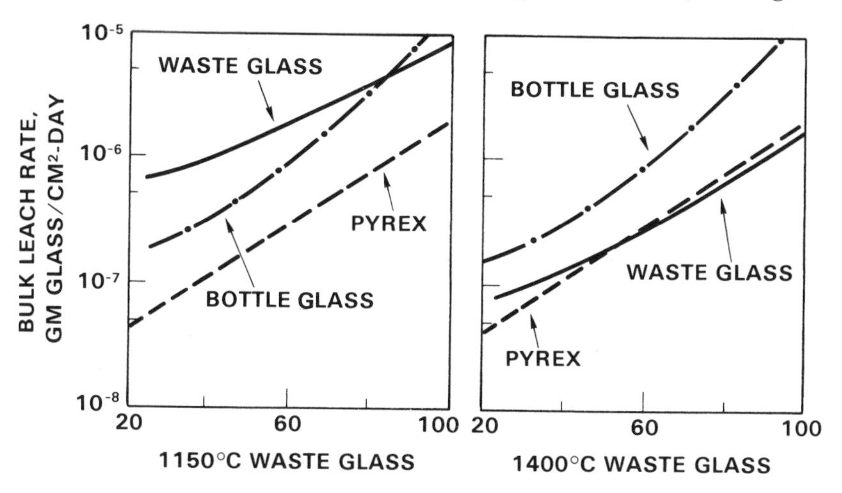

Figure 3. Comparative leach rates of waste glasses

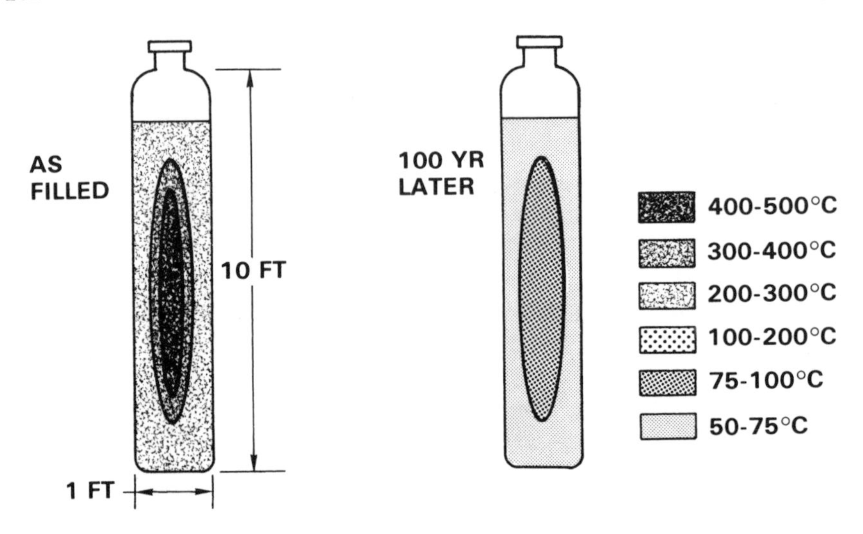

Figure 4. Temperature profile in representative HLW glass canisters in air storage

procedures will be designed to minimize thermal and physical shocks to the glass canisters.

Although for simplicity the word glass has been used in this paper to characterize the silicate waste fixation media, in reality they will be complicated mixtures of crystalline and glassy phases. This is caused by insolubility of a few constituents in the glassy matrix used. However, a more important reason is that the glass will be self-heating. The amount of self-heating is a function of the radioisotopic content of the glass and may vary over a wide range in the canisters of solidified commercial HLW, depending on such factors as fuel exposure and cooling time. A representative temperature profile in a typical canister of commercial waste glass is shown in Figure 4. The thermal conductivity of the waste glass is typically 1–1.2 W/m°C. This relatively low thermal conductivity causes the centerline temperature of the typical waste glass canister to approach 500°C initially and to remain above 100°C for about 100 years of storage. Although these temperatures are below the normal devitrification temperature range (600–900°C for representative silicate waste glasses), some devitrification may still occur because of the extremely long times involved. Devitrification refers to the spontaneous crystallization of certain constituents in the initially amorphous glass structure.

Consideration is being given to devitrifying purposely each of the waste glass canisters, since the devitrified form is more stable thermodynamically. Controlled devitrification is the technological basis of glass ceramics, which have seen rapidly spreading commercial application in the past 20 years. Properties of glass ceramics, which would be desirable in a HLW media vis-a-vis true glass, include:

- stability at higher temperatures
- greater thermal and mechanical shock resistance, and
- greater assurance of long-term stability.

The composition and thermal treatment of commercial glass ceramics are carefully controlled to achieve almost 100% conversion of the initial glass form to crystals of an essentially uniform size, a high proportion of which are less than one micron in size. Such careful control of composition and temperature will probably not be possible in HLW solidification, but the formation of quasi glass ceramics with many of the advantages of true glass ceramics may be possible. Investigations in this area are part of the PNL waste fixation program.

The high level of self-induced radiation in the waste glass causes elevated temperatures, and possibly several other effects. Preliminary analyses indicate that none of the other effects will significantly affect solidified waste management. Current investigations will verify these preliminary analyses experimentally. Some of the possible radiation effects to be studied are helium build-up, mechanical damage, and stored energy. Most radiation effects have some common characteristics: (a) they are caused mainly by the actinide content of the glass, and (b) they tend to be annealed out at elevated temperatures. Thus, it is only after the waste canisters have cooled (i.e., after most of the fission products have decayed out) that radiation effects of this type may become important. However, helium buildup from the decay of alpha particles is an exception; the possibility exists that helium may become concentrated in a band part way out from the centerline of very hot canisters. The glass inside the band, where helium diffusion was highest, would then be depleted in helium. Helium effects, plus radiation-induced lattice dislocations, can produce stresses in the glass which would ultimately lead to spontaneous cracking, or produce greater-than-normal cracking in the event the canister is thermally or mechanically shocked.

Stored energy refers to energy which can be secreted in solid materials caused by radiation-induced lattice displacements. Thermal energy can be released if the temperature of the material is raised above a certain level (which is unique for each material). The stored energy in waste glass is estimated not to exceed 200 cal/g, and analyses have indicated that the release of this amount of energy would not have serious effects on a waste storage or handling facility (*4, 5*).

The PNL program to study stored energy and other potential radiation effects is utilizing ^{244}Cm with a half-life of 18.1 yr. Use of this short-lived actinide permits accelerated radiation-effects studies. A series of glass specimens containing approximately 1 wt % curium is being prepared. The uniform distribution of curium achieved in these

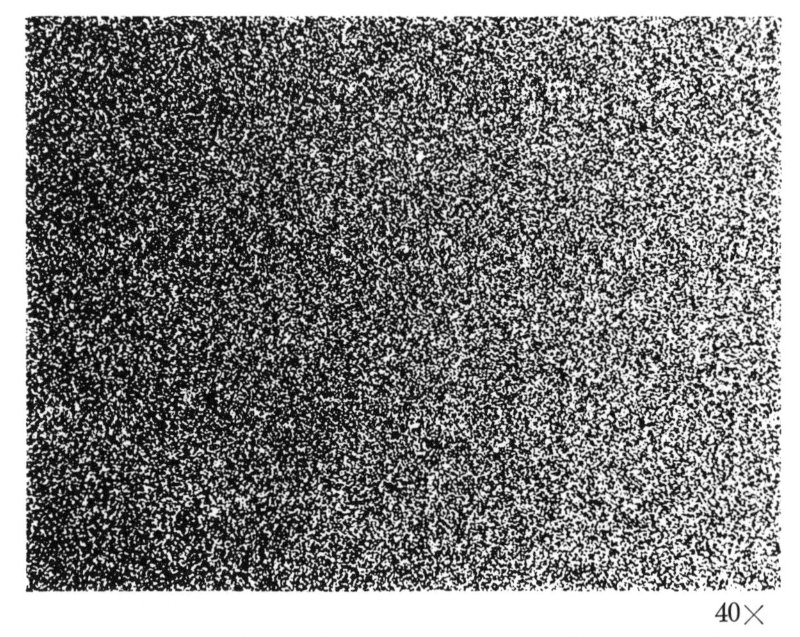

Figure 5. Autoradiograph of ^{244}Cm in zinc borosilicate waste glass

samples is shown by the autoradiograph in Figure 5. The number of alpha particles emitted in these specimens in one yr is equal to the number of alpha particles emitted in a typical HLW glass in its first 100 years. Measurements of stored energy, density changes, helium diffusivity, friability, and leach rate will be obtained on these specimens as a function of dose rate; however, no data are yet available. An additional stored energy program, in which simulated solidified wastes are being neutron irradiated, is being carried out in conjunction with ORNL.

The WSEP Borosilicate Glasses

Evidence already exists for the general thermal and radiation stability of borosilicate waste glass. As described earlier, three canisters of waste glass were made in the WSEP program. Samples of these products were obtained by core-drilling the canisters after 2–3 yr of storage. Leach rates of representative core-drilled samples from the canisters are compared with the leach rates of nonradioactive specimens in Figure 6. Note that two types of borosilicate glass were made in the WSEP program and that the leach rates of both types are considerably higher than those of the borosilicate glass investigated for the new PNL program; the earlier glass was formed in situ in its stainless steel canisters. Therefore, the temperature was maintained at 950°C or below to minimize corrosion of the canisters. The wall sections obtained during core drilling showed that no significant corrosion occurred.

The WSEP canister of rising-level borosilicate glass in particular was subjected to extreme storage conditions. While forming, the canister was generating 310 W/l. of radioactive decay energy. The canister was then held 1.8 yr in controlled storage with an average wall temperature of 425°C; during this period the centerline temperature of the 6-in.-diameter canister varied from 740° to 480°C. The initial leach rate was somewhat higher than obtained on nonradioactive simulated waste glass of the same composition, but the leach rate dropped within four weeks, becoming comparable with the nonradioactive comparison samples.

Two canisters of in-pot melting borosilicate glass were made in the WSEP program. Since it contained more silica and less B_2O_3, the in-pot melting borosilicate was a more leach-resistant glass than the rising-level glass. (It should be noted however that the rising-level glass contained almost 50 wt % fission products, illustrative of the very high-waste loading that can be achieved in borosilicate glass, albeit with some sacrifice in chemical durability.)

Core-drilled samples were taken from the in-pot melting glass canisters after the canisters had been stored almost three years and had received 1×10^{11} rad or more of radiation. Leach data from the core-drilled samples are shown to bracket the nonradioactive control data in Figure 6. Differences in leach rate of the magnitudes shown are within experimental deviation; these differences which are believed to be associated mainly with difficulty in determing surface areas accurately. The shape of the radioactive and nonradioactive leach rate curves for the

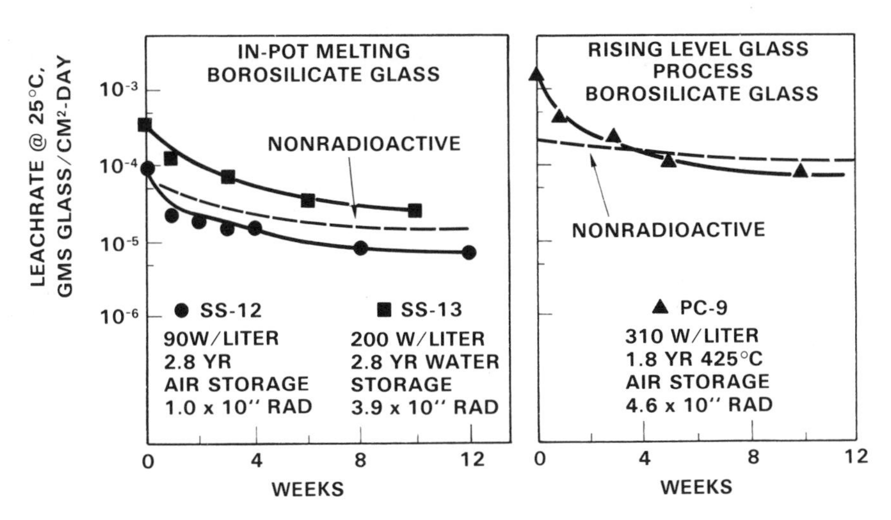

Figure 6. Effect of temperature and radiation on WSEP waste-glasses

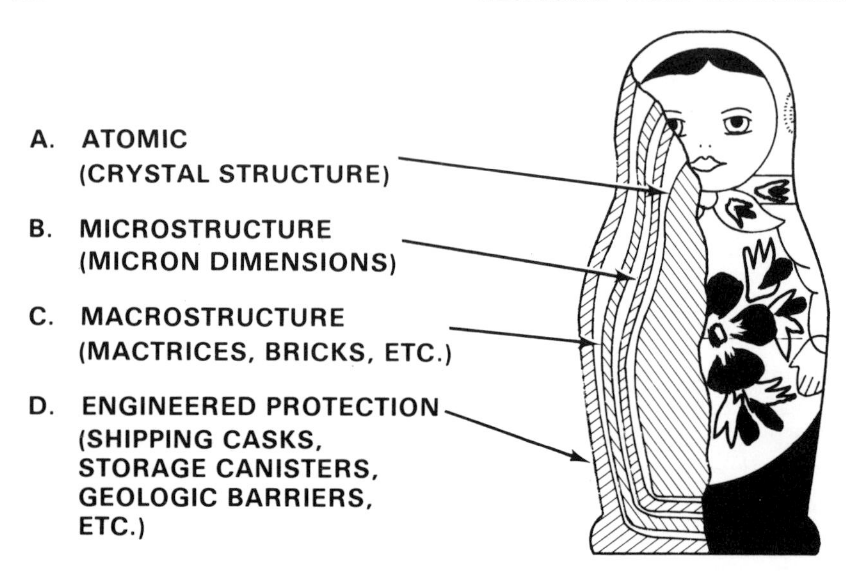

Figure 7. The "Russian Doll" concept

in-pot melting glass are similar, and no significant deterioration in the quality of this radioactive glass has occurred.

Advanced Waste Form Research

Although glass or other melt-formed ceramic forms are considered acceptable media for the permanent fixation of radioactive wastes, the AEC is sponsoring research to evaluate more advanced waste forms which may offer added increments of safety or economy in the future. This research is currently on a laboratory scale with nonradioactive materials.

The current glass processes produce radioactive glass encased in stainless steel, and the protection achieved can be augmented by the addition of more layers of inert material. Storage basins, shipping casks, etc. are common examples of this type of redundant protection, but the waste form itself can also be designed in layers. Rustum Roy, of Pennsylvania State University, has suggested that this concept be called the "Russian Doll" after the multilayered wooden toys of that name (*see* Figure 7). The goal of the "Russian Doll" approach is to tie up the individual radioactive atoms first, in the most stable configuration achievable; then various layers and coatings of different inert materials, each chosen for a certain specific property, can be added.

Two areas of the advanced waste form research, representative of the work currently being directed by the Nuclear Waste Technology Department at PNL, will be described briefly. One goal of the Penn State work (being done under a subcontract with PNL) is the development of inert, thermodynamically stable chemical forms for the individual radioactive atoms. This is a long-term fundamental research program involving the synthesis of tailor-made compounds of the fission products and the study of their properties. Hot pressing of either the tailor-made compounds or waste oxide calcine in inert matrix materials is also being investigated at Penn State. Promising results have been obtained using quartz as the matrix material (*6*).

Advanced waste form work is also being carried out in the Ceramics and Graphite Section at PNL, where high temperature gas-cooled reactor fuel technology is applied to waste solidification. Waste particles are coated with pyrolytic carbon followed by a cover coat of silicon carbide. These coated particles would then be placed in a matrix of inert material contained in a canister of yet another material.

Summary

The waste management research and development program at PNL may be summarized as consisting of (a) a mainline effort to develop near-term technology for the fixation of commercially-produced high-level wastes in silicate glass or ceramic forms, and (b) a comprehensive examination of potential advanced waste forms, a study which may offer further increments of safety or economy over the long term.

Literature Cited

1. Brill, R. H., "The Record of Time in Weathered Glass," *Archaeol.* (1961) **14,** 18–22.
2. Friedman, I., "Hydration Rind Dates Rhyolite Flows," *Science* (1968) **159,** 878–880.
3. McElroy, J. L., Schneider, K. J., Hartley, J. N., Mendel, J. E., Richardson, G. L., McKee, R. W., Blasewitz, A. G., "Waste Solidification Program Summary Report, Vol. II, Evaluation of WSEP High-Level Waste Solidification Process," **BNWL-1667,** July 1972.
4. Sonder, E., Nichols, J. P., Lindenbaum, S., Dillon, R. S., Blomeke, J. O., "Radioactive Waste Repository Project: Technical Status Reports for Period Ending September 30, 1971," **ORNL-4751,** December 1971, pp 187–202.
5. Laser, M., Merz, E., "The Question of the Energy Accumulation in Solids through Nuclear Irradiation in the Storage of Highly Radioactive Wastes," **JUL-766-CT,** Julich, Germany, 1971.
6. McCarthy, G. J., "Quartz Matrix Isolation of Radioactive Wastes," *J. Materials Science* (1973) **8,** pp 1358–1359.

RECEIVED November 27, 1974.

8

Fixation of Radioactive Waste by Hydrothermal Reactions with Clays

G. SCOTT BARNEY

Chemical Technology Laboratory,
Atlantic Richfield Hanford Co., Richland, Wash. 99352

A means of converting high-level radioactive waste containing sodium salts to solid, relatively insoluble aluminosilicate minerals, is described. Powdered aluminum silicate clays such as kaolin or bentonite are allowed to react with caustic, aqueous solutions of the waste at temperatures in the range of 30°–100°C. The reaction product from nitrate-containing wastes is a salt-filled, sodium aluminosilicate with the framework structure of cancrinite. Radioactive cesium and strontium are trapped in cancrinite along with sodium salts. The cancrinite crystals are spheres approximately 0.5 μm in diameter and are clustered. Leaching of the powdered product with distilled water at 30°C gives bulk leach rates in the range of 10^{-7} to 10^{-9} g of sample/cm^2-day based on BET surface areas.

High-level radioactive wastes from nuclear fuel reprocessing are presently being stored in underground tanks at several U.S. Atomic Energy Commission (now ERDA) installations. Some of these wastes are being solidified by evaporation to salt cakes in order to increase their safety during storage. In the future it may be necessary to convert these wastes into a form that will decrease the possibility of movement of radionuclides to the biosphere during long-term storage. Ideally, the waste should be in a form that would resist tank leakage, groundwater leaching, leaching by floods, wind erosion, and contamination spread by handling or sabotage.

General criteria for a desirable solid waste form suggested by the above safety considerations are as follows:

- Low leachability in water.
- Good chemical and physical stability toward heat and radiation.
- High mechanical strength.
- High thermal conductivity.
- Low volume.

The cost of production and storage of the waste form is also an important consideration.

Some aluminosilicate minerals appear to meet the above criteria rather well, especially with regard to low leachability and chemical and physical stability (*1, 2*). A low-temperature process for converting the wastes to aluminosilicates with low leachability has now been found (*3*). Aqueous waste solutions containing NaOH, $NaNO_3$, $NaNO_2$, $NaAlO_2$, mixed fission products, and minor amounts of other salts are mixed with powdered clays (kaolin, bentonite, halloysite, or dickite) and allowed to react at 30°–100°C to form small crystals of the mineral cancrinite. The sodium aluminosilicate crystal lattice of cancrinite contains large amounts of trapped salts and radioactive fission products. The process is applicable to caustic radioactive liquids such as neutralized Purex wastes or to salts or oxides produced by evaporation or calcination of these liquid wastes.

Summary and Conclusions

A low-temperature process for conversion of radioactive sodium salt wastes into solid, relatively insoluble, thermally stable sodium aluminosilicates is described. The reaction of the waste (in aqueous solution) with powdered clays such as kaolin, bentonite, halloysite, or dickite produces small crystals ($\simeq 0.5\ \mu m$) of cancrinite. Salts, including radioactive ones, are trapped in the cancrinite crystal lattice. The approximate chemical formula of the cancrinite produced is $2(NaAlSiO_4) \cdot x\ \text{salt} \cdot y\ H_2O$, with $x = 0.52$ and $y = 0.68$ when the entrapped salt is $NaNO_3$. The stoichiometry requires two moles of NaOH for each mole of cancrinite formed.

The reaction occurs at temperatures as low as 30°C. However there is a large temperature effect on the rate. The half-times for the reaction of kaolin with standard waste at 100°, 75°, and 50°C are, respectively, 1, 10, and 150 hr. The corresponding half-times for the bentonite reaction are 1, 9, and 60 hr.

The cancrinite product has a low leach rate. Values obtained by leaching with distilled water are in the range of 10^{-7} to 10^{-9} g of sample/cm^2-day based on BET surface areas of powdered samples. Diffusion of cesium ion from the product into the leachant is slower than for the sodium ion.

The melting range of cancrinite products is about 900°–1200°C. Zeolitic water is removed at 100°–300°C. At about 700°C the trapped $NaNO_2$ and $NaNO_3$ begin to decompose to Na_2O and nitrogen oxides. A change in crystal structure to nepheline occurs at 530°–595°C.

The thermal conductivity of cancrinite products ranges from 0.11 to 0.67 Btu/(hr) (ft^2) (°F/ft). These values appear to be high enough to

Table I. Analyses of Clay Reagents[a]

Component	Weight Percent of Component in Kaolin 1	Kaolin 2	Bentonite 1	Bentonite 2
SiO_2	46.20	43.97	54.74	63.40
Al_2O_3	38.06	41.93	19.25	18.66
Na_2O	0.067	0.220	3.51	2.72
Fe_2O_3	0.36		3.60	3.18
MgO	0.041		2.84	2.78
CaO	0.0007		0.14	1.43
H_2O	11.7	13.4	11.1	10.6
Total	96.4287	99.52	95.18	102.77

[a] Sources of clays: kaolin 1—J. T. Baker Chemical Co., kaolin 2—Georgia Kaolin Co., KCS-SD; bentonite 1—Robinson Laboratory, bentonite 2—Georgia Kaolin Co., MC-101.

prevent large temperature gradients in products made from Hanford stored waste.

The mechanical strength of the product can be improved by adding hardeners or binders. Various inorganic cements, including Portland cement, can be used to bind the cancrinite crystals to make a stronger product. Organic polymers can also be used. Binders appear to be most effective when the cancrinite is made from fired clays in which the clay structure has been destroyed. Various alternatives for the process are presented; these depend on whether or not a binder is necessary and at what point in the process a binder is added.

The volumes of the clay and waste solution are nearly additive. Therefore, the amount of clay necessary to form an acceptable product will determine the volume increase for the process. Fired clays appear to give the smallest volume increase (as low as 20% for the standard waste solution).

Experimental

Materials. Several sources of clays were used. Their compositions are given in Table I. Fired clays were prepared by heating the bentonite and kaolin at 600°–700°C for 24–48 hr.

Two synthetic waste solutions were used. The compositions of these solutions, along with that of an actual waste solution (from Hanford Tank 103-BY), are given in Table II.

Procedure. For most reactions, measured amounts of clay and waste solution were mixed thoroughly in polypropylene bottles. The bottles were then sealed and placed in a constant temperature oven (Blue M Electric Co.), shaker bath (Precision Scientific Co.), or boiling water bath for a specified time. For kinetic experiments the clay and waste solutions were allowed to reach the desired temperature overnight. Waste solution was measured and added to the clay using a preheated plastic syringe. Samples were periodically removed from the oven and filtered hot with a 5-μm Teflon millipore filter in a Nalgene plastic filter holder. The filtrate was collected, and the solids were washed five times with 10-ml portions of distilled water. The solids were then air-dried and weighed.

Leach rates were measured by shaking 1.0 g of powdered sample with 200 ml of distilled water in a constant temperature bath set at 30°C. The mixtures were filtered periodically, and the solids were contacted with fresh 200-ml portions of distilled water. The leachates were then analyzed for various cations.

Analyses. Elemental analyses of the solid products were obtained by fusing the sample with lithium metaborate at 1000°C and then dissolving the melt in 10% HCl. The resulting solutions were diluted to 100.0 ml and were analyzed by atomic absorption (Varian techtron model 385). Cations in leachates were also measured with this instrument.

Crystalline products were identified by x-ray diffraction using a General Electric XRD-5 powder diffractometer. Thermal analyses were performed with Mettler recording vacuum thermoanalyzer. Surface areas of powdered samples were measured using the BET method. Thermal conductivities of clay-waste mixtures and reaction products were measured according to the method of Underwood and McTaggart (*4*). Hardness of several products was measured with a Soiltest penetrometer, model CL-700. Scanning electron microscope photographs of the crystalline products were obtained from Battelle Pacific Northwest Laboratories. The volume of "free liquid" left after a reaction was measured by filtering the cooled reaction mixture through an 0.5-μm Solvinert millipore filter at 100 psi until no more filtrate could be obtained.

Results and Discussion

Formation and Identification of Product. The reaction of kaolin with basic solutions of various salts has been studied previously by Barrer and co-workers (*5-8*). They obtained salt-filled cancrinites from the reaction of kaolin with NaOH solutions containing $NaNO_3$, Na_2SO_4, Na_2CrO_4, and several other salts. The salts are incorporated into the 11-hedral cancrinite

Table II. Composition of Synthetic and Actual Waste Solutions

	Molar Concentration of Components in		
Component	*Synthetic Waste 1*	*Standard Synthetic Waste*	*Actual Waste*[a]
NaOH	5.0	3.6	5.2
$NaNO_3$	2.0	3.4	1.9
$NaNO_2$	1.0	2.1	2.0
$NaAlO_2$	1.8	1.8	2.7
Na_2CO_3	0.1	0	–
Na_2SO_4	0.05	0	–
H_2O	–	40	–

[a] From tank 103-BY, also containing 4.60×10^5 μCi/liter ^{137}Cs.

Table III. Comparison of X-Ray Diffraction Patterns of Cancrinites and Clay Reaction Product

Natural Cancrinite[a]		*Synthetic Cancrinite*[b]		*Natrodavyne*[c]		*Clay Reaction Product*	
dA	I/I_1	*dA*	I/I_1	*dA*	I/I_1	*dA*	I/I_1
11.0	30						
6.32	60			6.38	70	6.27	50
5.47	10						
4.64	80	4.70	80	4.64	10	4.68	60
4.13	30						
3.75	10						
3.65	50	3.67	80	3.68	100	3.63	80
3.22	100	3.25	100	3.21	10	3.23	100
3.03	30						
2.97	30						
		2.82	16	2.85	25		
2.73	60	2.76	35			2.73	35
2.61	50	2.63	30	2.60	35	2.61	30
2.56	60	2.57	16			2.58	35

[a] ASTM powder diffraction file, card No. 20-257.
[b] ASTM powder diffraction file, card No. 15-734.
[c] ASTM powder diffraction file, card No. 15-469.

cages and into the wide channels enclosed by these (9). The entrained salts could not be released except by decomposition of the aluminosilicate with mineral acid. The present work shows that radioactive salts of cesium and strontium can also be entrapped in the cancrinite products.

IDENTIFICATION OF PRODUCTS. The crystalline product obtained from reaction of the clays with either synthetic or actual wastes was identified as cancrinite. A comparison of the x-ray diffraction patterns of the clay reaction product with three other types of cancrinites is given in Table III. The match with natural cancrinite and synthetic cancrinite is very good.

The presence of small amounts of fission products in the waste does not affect the formation of cancrinite. Reactions of bentonite and synthetic waste spiked with large amounts of Cs^+ and Sr^{2+} gave only cancrinite, even when the concentrations were many times greater than those expected in reprocessing wastes. At yet higher concentrations of Cs^+ and Sr^{2+} other products were formed, as shown in Table IV.

The clay/waste solution ratio appears to have no effect on the identity of the crystalline aluminosilicate product. Reactions over wide ranges of clay/synthetic waste ratio gave only cancrinite for those clays studied (bentonite, kaolin, halloysite, and dickite). Also, the reaction of actual waste solution (tank 103-BY) with bentonite over the range 0.05–2.0 g of

Table IV. Effect of Cs^+ and Sr^{2+} Concentration in Synthetic Waste on Reaction Products[a]

Sample	*Molarity of Cs^+ or Sr^{2+} in Synthetic Waste 1*	*Products*
122	0.37M Cs^+	Cancrinite
123	0.75M Cs^+	Cancrinite + unknown
124	1.50M Cs^+	Cancrinite + unknown
125	2.25M Cs^+	Cancrinite + pollucite
126	0.10M Sr^{2+}	Cancrinite
127	0.38M Sr^{2+}	Cancrinite + $Sr_3Al_2O_6 \cdot 6H_2O$
128	1.00M Sr^{2+}	Cancrinite + $Sr_3Al_2O_6 \cdot 6H_2O + SrCO_3$
129	1.30M Sr^{2+}	Cancrinite + $Sr_3Al_2O_6 \cdot 6H_2O + SrCO_3$

[a] 20 ml of solution reacted with 5.0 g of bentonite at 100°C for 10 days.

clay/ml solution gave only cancrinite (plus unreacted bentonite at the higher ratios).

Cancrinite was also formed by reacting ignited kaolin and bentonite with synthetic waste. All the interlayer and hydroxyl water was removed from these clays before using by heating at 600°–700°C for 24–48 hr. The products from these reactions will presumably have a smaller volume because of the removal of the water.

STOICHIOMETRY. The stoichiometry of the formation of cancrinite from kaolin was studied. The amount of NaOH which reacts with a given amount of clay was determined by mixing the clay with a solution containing various concentrations of NaOH and a constant concentration of $NaNO_3$ (necessary to form the cancrinite). Essentially all of the NaOH in the solutions reacted during cancrinite formation. The mole ratios of NaOH, $NaNO_3$, and kaolin reacted are given in Table V, based on the formula $Al_2Si_2O_7 \cdot 2H_2O$ for kaolin (MW = 259).

Table V. Mole Ratios of Reactants Used[a]

[NaOH], M	*Moles NaOH / Moles Kaolin Reacted*	*Moles $NaNO_3$ / Moles Kaolin Reacted*
1.0	1.7	0.62
2.0	1.7	0.47
3.0	2.1	0.57
4.0	2.0	0.47
5.0	2.1	0.45
	average = 1.9	average = 0.52

[a] 5 ml of 2.0M $NaNO_3$ at the given NaOH concentration reacted with 5.0 g of kaolin 1 for 11 days at 100°C.

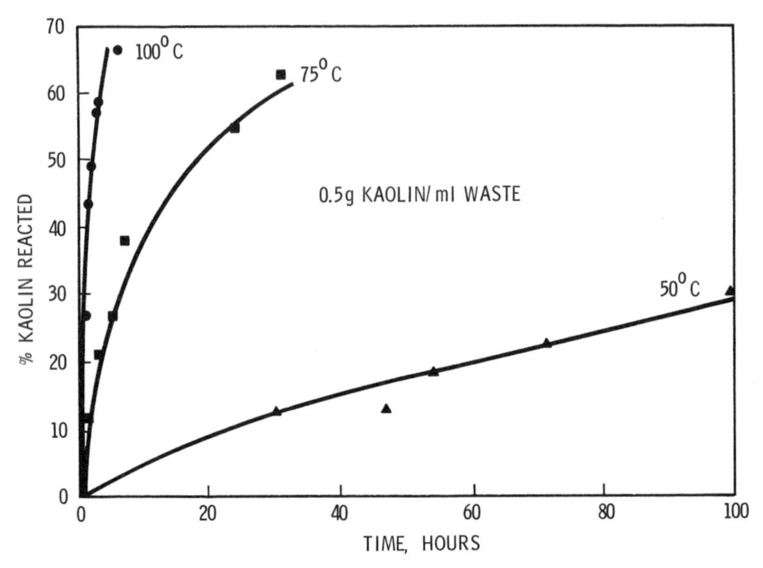

Figure 1. Effect of temperature on reaction rate (kaolin-waste reaction)

The amount of zeolitic water in the cancrinite was $\simeq$ 0.68 mole per mole of cancrinite formed. These observations can be summarized by Reaction 1:

$$\underset{\text{(kaolinite)}}{Al_2Si_2O_7 \cdot 2H_2O} + 2NaOH + 0.52NaNO_3 + 0.68H_2O \rightarrow \tag{1}$$

$$\underset{\text{(cancrinite)}}{2(NaAlSiO_4) \cdot 0.52NaNO_3 \cdot 0.68H_2O} + 3H_2O$$

The cancrinite composition is very near that reported by Barrer (*8*), $2(NaAlSiO_4) \cdot 0.65NaNO_3 \cdot 0.60H_2O$.

Similar equations can be written for the reaction of bentonite (which consists mainly of the mineral montmorillonite) with the caustic $NaNO_3$ solutions. Assuming that the formula of the cancrinite is the same as above,

$$\underset{\text{(montmorillonite)}}{Al_2Si_4O_{11} \cdot H_2O} + 2NaOH + 0.52NaNO_3 + 2.68H_2O \rightarrow \tag{2}$$

$$\underset{\text{(cancrinite)}}{2(NaAlSiO_4) \cdot 0.52NaNO_3 \cdot 0.68H_2O} + 2H_4SiO_4$$

or, in the presence of $NaAlO_2$,

$$\underset{\text{(montmorillonite)}}{Al_2Si_4O_{11} \cdot H_2O} + 2NaAlO_2 + 2NaOH + 1.04NaNO_3 + 1.36H_2O \rightarrow \tag{3}$$

$$\underset{\text{(cancrinite)}}{2[2(NaAlSiO_4) \cdot 0.52NaNO_3 \cdot 0.68H_2O]} + 2H_2O$$

Other salts, in addition to $NaNO_3$, can be trapped in the cancrinite lattice.

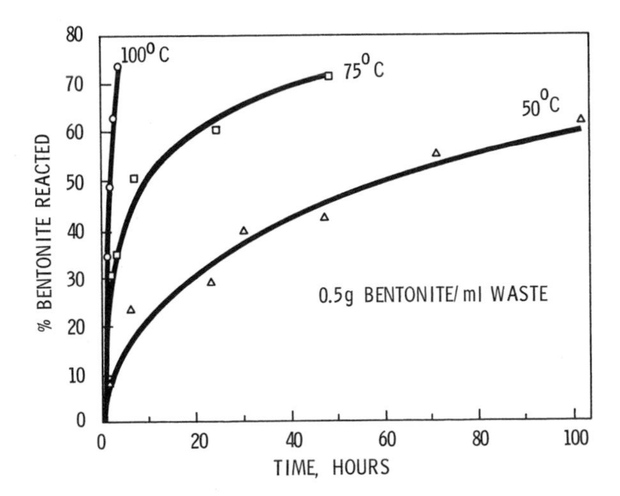

Figure 2. Effect of temperature on reaction rate (bentonite-waste reaction)

KINETICS. The kinetics of the reactions of bentonite 2 and kaolin 2 clays with standard synthetic waste solution were studied at three different temperatures: 50°, 75°, and 100°C. The extent of the kaolin reaction was measured by determining (using thermal analysis) the amount of hydroxyl water associated with the unreacted kaolin in the solid product. The amount of unreacted bentonite was estimated by measuring the area of the endotherm attributed to loss of hydroxyl water from the bentonite. Reaction rate curves for the kaolin clay–waste reaction and bentonite clay–waste reaction at the above temperatures are shown in Figures 1 and 2, respectively. The great effect of temperature on the rate of reaction is apparent from these curves. It should be pointed out that there was an excess of clay in these experiments so that the maximum amount of clay reacted could be only $\simeq 80\%$. Since the standard waste solution contains only 3.6*M* NaOH, it is expected that higher concentrations of hydroxide will increase the reaction rate.

During the kinetic experiments the solid and liquid phases were analyzed for Na^+, Al^{3+}, Si^{4+}, NO_3^-, NO_2^-, and clay hydroxyl water. The results, given as the weight percent of a given component in the solid phase as a function of time, are shown in Figures 3 and 4. Insight into the mechanism and rate of reaction can be obtained from these plots. Figure 3 shows that for the kaolin–waste reaction a substantial amount of aluminum remains in solution while almost all the silicon is in the solid phase. Figure 4 shows that the opposite is true for the bentonite–waste reaction. The driving force for the formation of a product with an Al/Si mole ratio of one (as is the case for cancrinite) is apparent from Table VI.

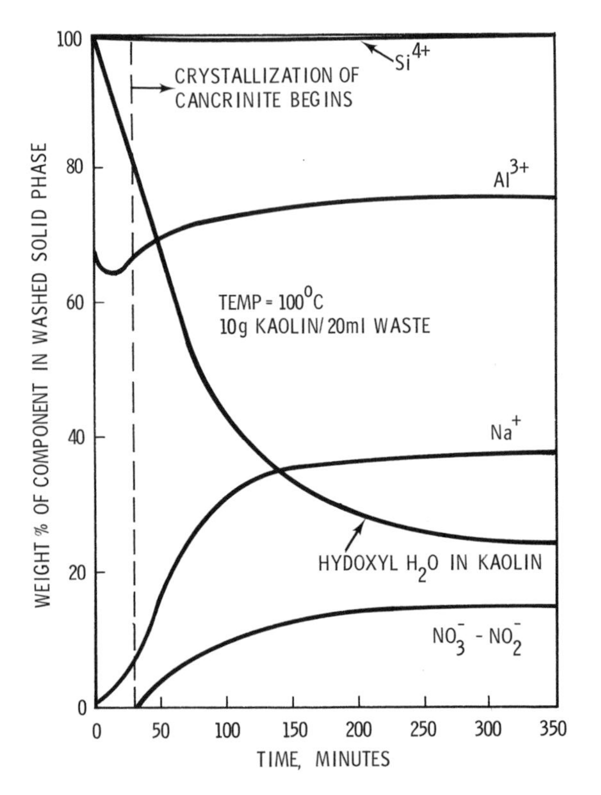

Figure 3. Solid phase composition vs. *time (kaolin-wate reaction)*

Table VI. Aluminum/Silicon Mole Ratios for Reaction System

Reaction	*Al/Si Mole Ratio*	
	Liquid + Solid Phase	*Solid Product*
Kaolin–waste	1.5	1.1
Bentonite–waste	0.82	0.96

The component (Al or Si) which is in excess for cancrinite formation remains in solution. The spikes in the Na^+ and NO_3^-–NO_2^- curves in Figure 4 at about 15 min are probably the result of initial ion exchange of the sodium salts with bentonite.

X-ray diffraction of the products from the above experiments show crystalline cancrinite in each sample from the bentonite reaction (the first sample was taken at 15 min). However, no cancrinite is observed in the kaolin reaction until 60 min has elapsed. The induction period for cancrinite crystal growth is, therefore, shorter with bentonite as the reactant. This may be associated with the rate of dissolution of the clays.

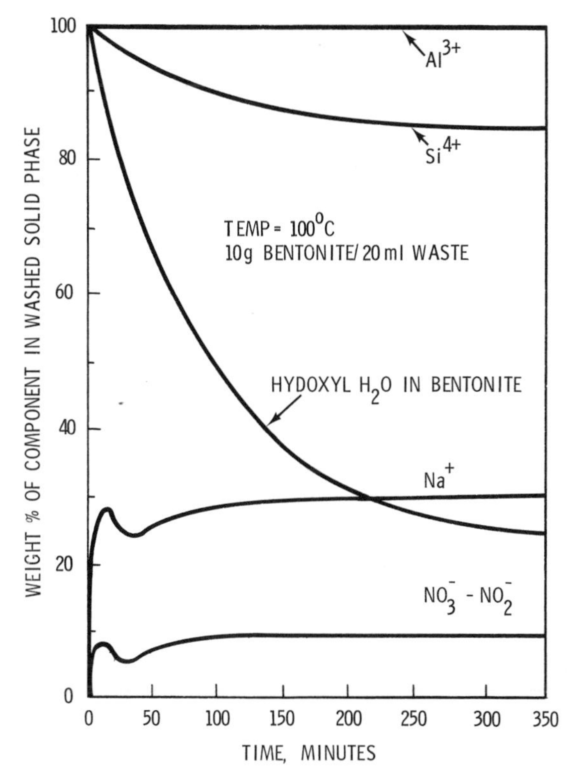

Figure 4. Solid phase composition vs. *time (bentonite-waste reaction)*

FIXATION OF CESIUM AND STRONTIUM. We mixed actual waste solution (103-BY) with bentonite 1 to determine the ability of the clay to fix cesium. The clay–waste mixtures were heated at 100°C for seven days. The percent of ^{137}Cs fixed in the product was measured as a function of the clay/liquid waste ratio. The results are plotted in Figure 5. The curve shows that a minimum of 0.75 g of bentonite 1 per ml of waste solution is needed to fix the ^{137}Cs.

Cesium was the only measurable radioactive component in the 103-BY waste. Fixation of other fission products could not, therefore, be studied using this waste. However, reactions of clays with synthetic waste spiked with strontium indicate that essentially all of the strontium is fixed in the solid product. Additional work with strontium and other major radioactive fission products is necessary.

Chemical and Physical Properties of Products

LEACH RATES. Leach rates for the cancrinite products have been determined by contacting either powdered or massive samples with distilled

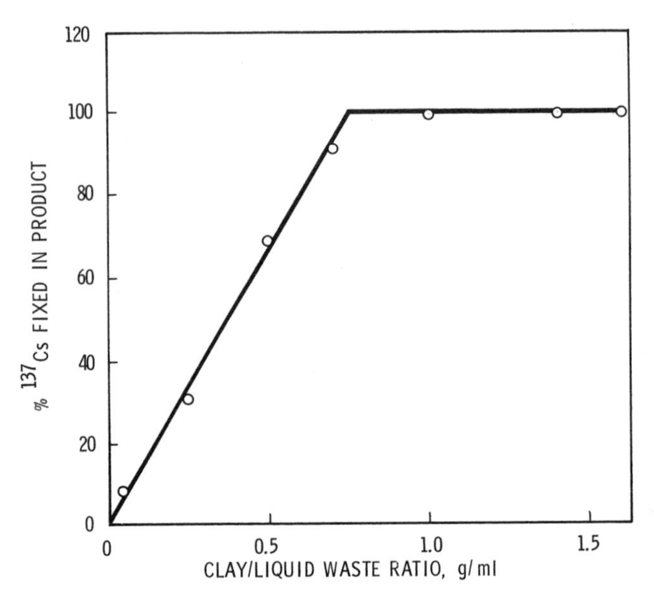

Figure 5. Fixation of ^{137}Cs using various clay/liquid waste ratios (reaction of bentonite 1 with actual 103-BY waste)

water. Since leach rates generally decrease with time, measurements were made over several weeks. Figure 6 gives leach rate curves for products obtained using various clays and waste solutions. Leach rates for the two products obtained from synthetic waste were calculated from the fraction of Na^+ leached. For the radioactive product the fraction of $^{137}Cs^+$ leached was used. The lower rate for the radioactive product is probably the result of slower diffusion of Cs^+ compared with Na^+. The calculation does not take into account diffusion of the measured ions since the assumption of congruent dissolution is made. Since diffusion of other components of the cancrinite product (such as Si and Al) will be slow compared to that of Na^+ and Cs^+, the actual bulk leach rate will be less than the calculated values. Table VII gives a summary of leach rate values obtained for powdered and massive samples. Some were washed with distilled water to remove any unreacted salts. Washing the cancrinite product lowers the initial leach rate. Because of different methods of surface area measurement, higher leach rates were calculated for massive samples than for powdered samples. Geometric surface areas were used for massive samples and BET surface areas were used for powdered samples. Since the massive samples are somewhat porous, BET surface areas are much larger than geometric surface areas.

Table VII. Leach Rate[a] for Cancrinite Products

	Leach Rate Range, g/cm²-day	
Sample Form	*Washed*	*Unwashed*
Powder	10^{-7}–10^{-9}	—
Massive	10^{-3}–10^{-5}	10^{-2}–10^{-4}

[a] Bulk leach rate $= \frac{\text{(fraction of ion leached)}}{\text{(surface area)}} \frac{\text{(sample weight)}}{\text{(time)}}$

THERMAL STABILITY. The melting points of cancrinite products fall in the range 900°–1200°C. Several changes in the composition of the products occur before the melting point is reached, however. The weight loss curve and differential temperature (DTA) curve for a typical product made from kaolin are given in Figure 7. The initial weight loss and associated endotherm are caused by loss of zeolitic water. Since there is some unreacted kaolin present in this product, the loss of hydroxyl water from the clay is observed at 450°–650°C. At about 700°C the trapped $NaNO_3$ and $NaNO_2$ begin to decompose to Na_2O and nitrogen oxides. There is a large exotherm and weight loss due to this decomposition.

A change in crystal structure of the product occurs at high tempera-

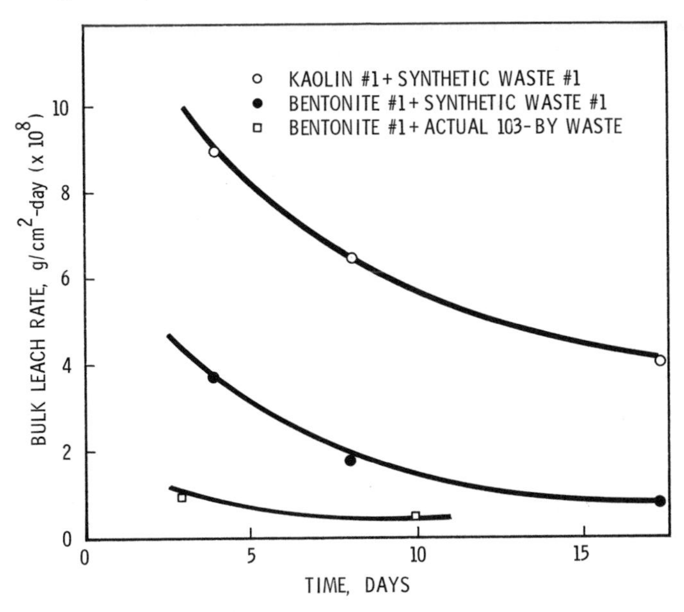

Figure 6. Bulk leach rate of cancrinite product calculated by the equation

$$\text{Bulk leach rate} = \frac{\text{(fraction of ion leached)}}{\text{(surface area)}} \frac{\text{(sample weight)}}{\text{(time)}}$$

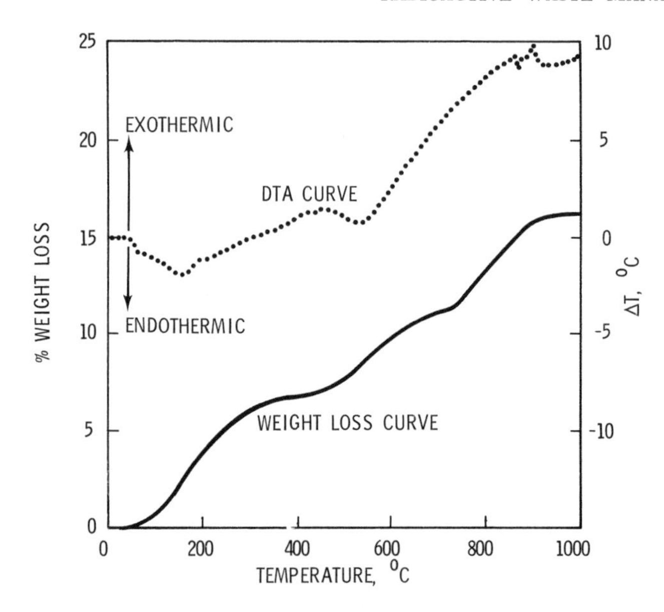

Figure 7. Thermal analysis of typical cancrinite sample; conditions: 500.3 mg sample in platinum crucible; heating rate 10°C/min in flowing air

ture. Samples heated from 800°–1000°C give an x-ray diffraction pattern corresponding to nepheline. According to Barrar (*8*), sodium nitrate cancrinite decomposes to nepheline somewhere in the range 530°–594°C.

THERMAL CONDUCTIVITY. Thermal conductivity measurements were made on clay–waste slurries and on the dried cancrinite product. These measurements represent the possible extremes. The values for the actual product, depending on the process, are somewhere between these extremes. The commercial clays, KCS (kaolin 2) and MC-101 (bentonite 2) were used with standard synthetic waste to make the clay–waste slurries and the dried cancrinite product. The results of these measurements are given in Table VIII. These values are near those of dried salt cake and are high enough so that large temperature gradients should not occur in the cancrinite product made from stored Hanford wastes.

HARDNESS. If the solid waste form is to be handled or transported, it must be strong enough to prevent chipping and dusting of small particles from the bulk. Pure cancrinite is quite hard (5–6 on Mohs' scale). However the product of clay–waste reactions is a mixture of small cancrinite crystals and unreacted kaolin (2.0–2.5 on Mohs' scales). The hardness of this product is, therefore, limited to that of kaolin. Also, the cancrinite product containing excess kaolin clay as a binder has been observed to soften somewhat when covered with water. If the cancrinite product contains excess bentonite clay, the product will swell and crumble when covered with water. To make a mechanically stronger product, several hardeners and binders have been tested. These can be added either to the reaction mixture or the cancrinite product.

Experiments thus far have shown that the addition of CaO, silica gel, or mixtures of both will give a harder product. These additions were made to the reaction mixture (10.0 g kaolin 2 plus 10 ml standard synthetic waste), and the samples kept at 75°C for 48 hr. The cooled, damp products were then measured for hardness with penetrometer (Soiltest model CL-700 penetrometer—Soiltest, Inc.). The results are given in Table IX. The greatest increase in hardness results from the addition of a mixture of CaO and $SiO_2 \cdot x\ H_2O$.

A dramatic increase in the hardness of dried cancrinite products occurs after treatment with tetraethyl silicate (TES). Soaking the cancrinite–clay mixture in TES for several hours, either at room temperature or at 100°C, gives a product which is difficult to scratch and which is more water resistant. Samples treated with TES do not appear to soften even when soaked in water for months.

Cancrinite crystals were used as aggregate in Portland type III cement mixtures to bind the crystals. Hard products were obtained only when the cancrinite was prepared from fired clays. It appears that any unreacted clay in the cement mixture will cause the cement to crumble when air dried. This may be due to shrinkage of the clay upon drying.

Other organic and inorganic binders have also been tested. Polymerization of styrene or methyl methacrylate in cancrinite–monomer mixtures produces hard, rugged products. Sodium silicate solution mixed with powdered silica also appears to be an effective binder for the cancrinite.

Table VIII. Thermal Conductivity of Clay–Waste Mixtures and Cancrinite Product

Sample	*Temperature, °C*	*Thermal Conductivity, Btu/(hr)(ft²)(°F/ft)*
Kaolin slurry		
1.0 g/ml mixture	24.9	0.670
Bentonite slurry		
1.0 g/ml mixture	24.7	0.458
Kaolin cancrinite		
product dry	25.3	0.140
product dry	44.0	0.158
product dry	74.7	0.161
Bentonite cancrinite		
product dry	26.3	0.114
product dry	56.4	0.185
product dry	69.0	0.175

Table IX. Effects of CaO and $SiO_2 \cdot x\ H_2O$ on the Hardness of Cancrinite Product

Mixture	*Weight CaO, g*	*Weight $SiO_2 \cdot xH_2O$, g*	*Hardness*[a] *kg/cm^2*
1	0	0	37
2	0	0.5	22
3	0	1.0	35
4	0	2.0	$>$ 45
5	0.5	0	45
6	1.0	0	$>$ 45
7	2.0	0	$>$ 45
8	3.0	0	$>$ 45
9	2.0	0.5	$>$ 45
10[b]	2.0	2.0	$>>$ 45[b]
11	1.0	2.0	$>>$ 45
12	0.5	2.0	$>$ 45

[a] The pressure necessary to drive the penetrometer probe 0.25 in. into the product.
[b] Hardest product tested.

VOLUME INCREASE. The volumes of the clay and waste solution are approximately additive. Since the density of both kaolin and bentonite is about 2.6 g/ml, each gram of clay added results in a volume increase of 0.38 ml. The amount of clay necessary to form a solid product depends on the type of clay used and the composition of the waste solution reacted.

The minimum amount of clay necessary to solidify the standard waste solution completely was measured for several clays. A solid product is defined as one that can be filtered at 100 psi using a pressure filter, with no resulting liquid. The results of these measurements for solidification of the standard waste solution are given in Table X. These results indicate that less volume is necessary if kaolin is used. Reactions with fired clays have resulted in volume increases as low as 20% (*10*).

The volume increase for salt cake wastes has not yet been determined. It does seem reasonable to assume that this volume increase will be greater because of the higher concentration of salts. There are many unknown variables that can influence the volume (bulk density of salt cakes, composition, water content, etc.).

SIZE AND SHAPE OF CANCRINITE CRYSTALS. A series of scanning electron microscope (SEM) photographs was made of the products of standard waste with kaolin 1 and bentonite 1. These photographs are shown in Figures 8 and 9. The cancrinite appears to be a mass of small, agglomerated spheres. The spherulites vary in diameter from about 0.5–5 μm. Increasing the initial waste solution/clay ratio in the reaction mixture gave larger spherulites.

DESCRIPTION OF POSSIBLE PROCESSES. It has been shown that radioactive wastes containing sodium salts can be converted to cancrinite, a solid,

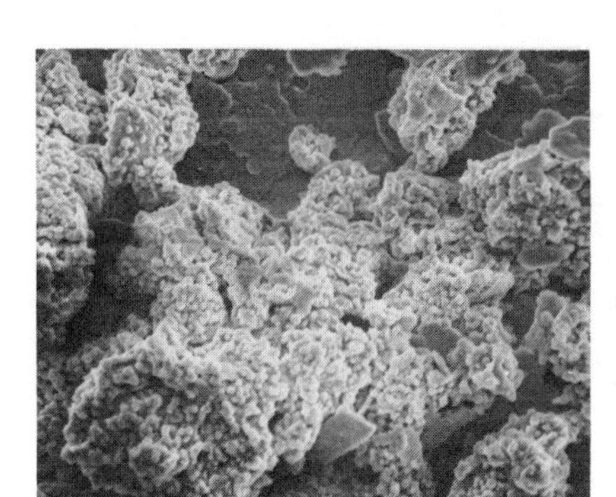
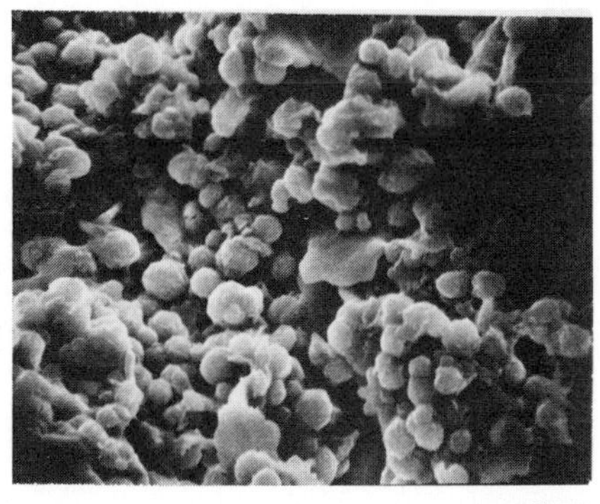

500X **250X**

Figure 8. SEM photographs of bentonite product

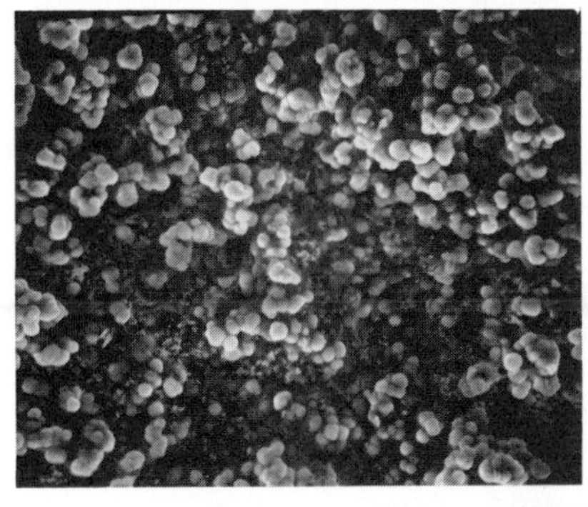
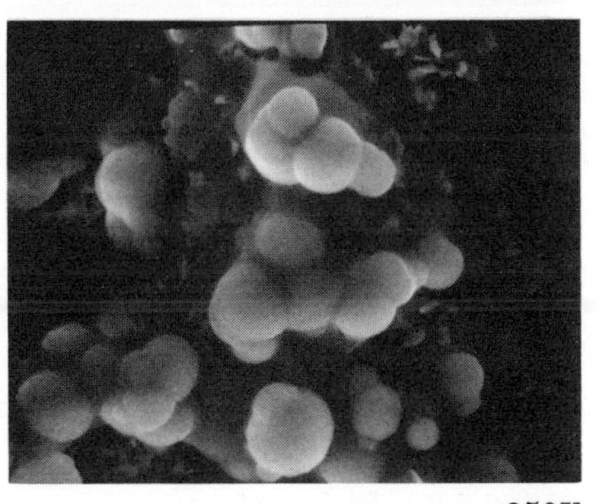

500X **250X**

Figure 9. SEM photographs of kaolin product

Table X. Minimum Amount of Clay Necessary to Solidify the Standard Waste Solution

Type of Clay	*Minimum Clay/ Liquid Waste Ratio, g/ml*	*% Volume Increase for Solid Product*[a]
Kaolin 2	0.70	26
Bentonite 1	0.84	32
Bentonite 2	1.30	49

[a] Compared with volume of standard waste solution.

relatively insoluble, thermally stable aluminosilicate mineral. Another possible product is nepheline which can be made by firing the cancrinite. A conceptual flow diagram showing several possible alternative treatments and products is given in Figure 10. Alternative 4 is the simplest and least expensive route to a cancrinite product. This product may not,

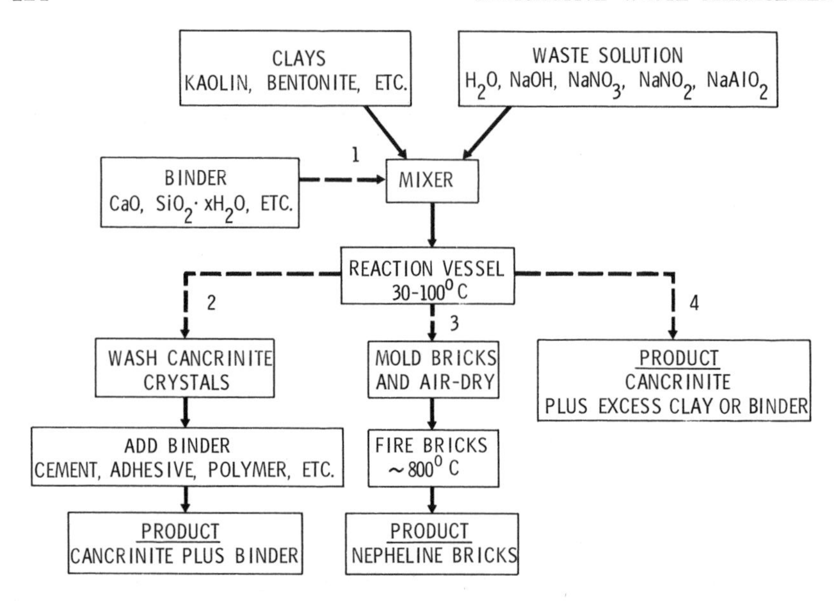

Figure 10. A conceptual flow diagram for the conversion of radioactive wastes to aluminosilicate minerals

however, meet the requirement of high mechanical strength. Alternative routes 1 + 4, 2, and 3 will give a more rugged product but will be more expensive. It may not be possible to add certain types of binders (polymers, cements, asphalt, etc.) to the reaction mixture. In this case alternative 2 must be used. The washing step was included to remove any unreacted salts. This step may not be necessary, depending on the kind of binder used.

Acknowledgments

The author expresses appreciation to W. I. Winters and W. L. Louk for atomic absorption analyses, to L. H. Taylor and S. G. Metcalf for x-ray diffraction analyses, and to D. G. Bouse for thermal conductivity measurements. Experimental assistance by J. P. Martelli is gratefully acknowledged.

Literature Cited

1. Patrick, W. A., "Use of Artificial Clays in Removal and Fixation of Radioactive Nuclides," Sanitary Engineering Conference, Cincinnati, 1955, *USAEC Rep.* (1956) **TID-7517,** 368.
2. Correns, C. W., "Chemical Disintegration of Silicates," *Naturwissenshaften* (1940) **28.**
3. Barney, G. S. Brownell, L. E., "The Fixation of Radioactive Wastes in Cancrinite for Safe, Long-Term Storage," Report of Invention ARH-IR-195, Atlantic Richfield Hanford Co., Richland, Wash. (1973).

4. Underwood, W. M., McTaggart, R. B., "The Thermal Conductivity of Several Plastics, Measured by an Unsteady State Method," *Natl. Heat Transfer Conf., 3rd, ASME-TIChE, Storrs, Conn. Aug. 9–12, 1959.*
5. Barrer, R. M., Cole, J. F., Sticher, H., "Chemistry of Soil Minerals. Part V. Low Temperature Hydrothermal Transformation of Kaolinite," *J. Chem. Soc. A* (1968) 2475.
6. Barrer, R. M., Marcilly, C., "Hydrothermal Chemistry of Silicates. Part XV. Synthesis and Nature of Some Salt Bearing Aluminosilicates," *J. Chem. Soc. A* (1970) 2735.
7. Barrer, R. M., Cole, J. F., "Chemistry of Soil Minerals. Part VI. Salt Entrainment by Sodalite and Cancrinite During Their Synthesis," *J. Chem. Soc.* (1970) 1516.
8. Barrer, R. M., Cole, J. F., Villiger, H., "Chemistry of Soil Minerals. Part VII. Synthesis, Properties and Crystal Structures of Salt-Filled Cancrinites," *J. Chem. Soc. A* (1970) 1523.
9. Jarchow, Z., *Krist* (1965) **122**, 407.
10. Delegard, C. H., personal communication, Atlantic Richfield Hanford Co. (1974).

RECEIVED November 27, 1974.

9

The Distribution of Plutonium in a Rock Containment Environment

S. FRIED, A. M. FRIEDMAN,[1] and R. WEEBER[2]

Chemistry Division, Argonne National Laboratory, Argonne, Ill. 60439

Studies of the migration of Pu in limestones and basalts indicate that the absorption coefficients are dependent on the types and amounts of other ions present in the solution. Migration coefficients were measured for flow along the surface of fissures and through the porous stone. At least two chemical forms of Pu were present in neutral solutions and one of these, presumably a polymerized Pu oxide, migrated 10 times faster than the other form.

The increasing amounts of radioactive waste material accumulating from reactor operations makes the disposal and safekeeping of this material vitally important. This is especially true for those nuclides of long half-life.

Regardless of the technical details of a particular disposal method, it is obvious that in a depository the ultimate secondary container must be rock strata. It is necessary to consider the rock strata not only as barriers but also as possible conduits for ground waters and as media for the dispersal of radionuclides and their ultimate incorporation into the biosphere. Studies have been undertaken to determine the behavior of long-lived radionuclides, particularly Pu and Am ($T_{\frac{1}{2}}$ 24,000 years and 450 years, respectively) in rock strata.

Experimental

The experiments were of two types. In the first set, studies were made of the migration of solutions of Pu through cores of Niagara limestone and basalts. A high pressure chromatographic absorption appa-

[1] and College of Environmental and Applied Sciences, Governor's State University, Park Forest, Ill. 60466.

[2] Thesis parts student, Governor's State University, Park Forest, Ill. 60466.

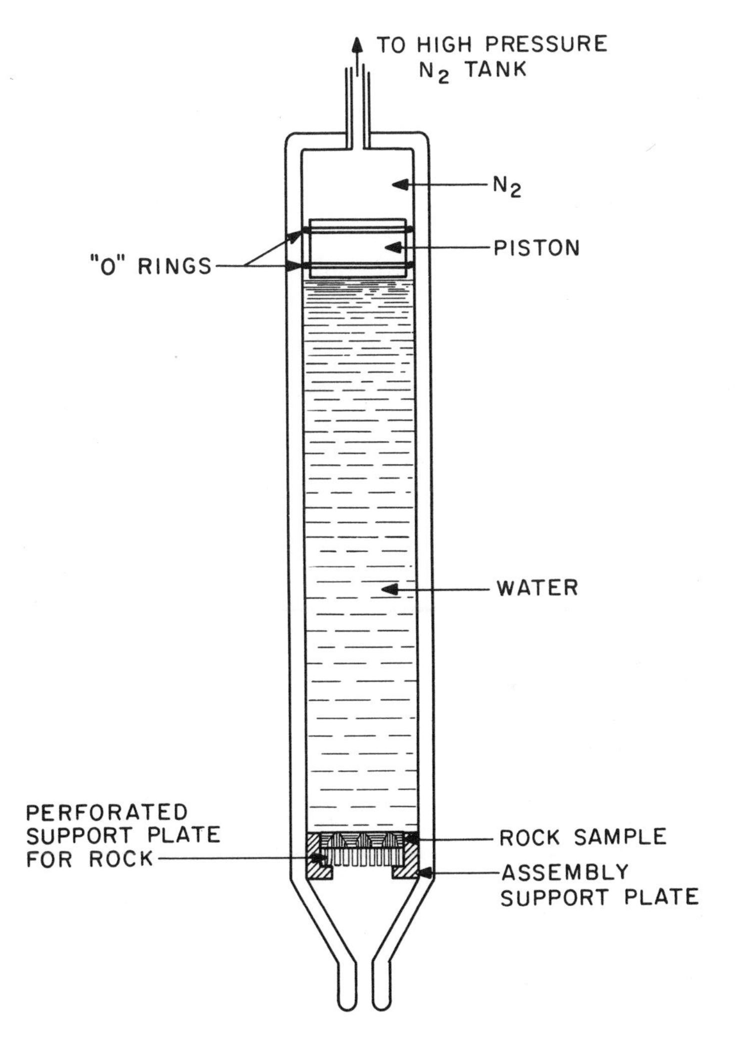

Figure 1. Schematic diagram of the high pressure chromatographic column. The apparatus was constructed of stainless steel.

ratus was constructed and is shown diagrammatically in Figure 1. The function of the piston is to exert pressure on the solution while isolating it from the pressurizing gas.

At the start of an experiment a small amount of $^{238}Pu(NO_3)_4$ tracer in neutral aqueous solution was placed on the surface of the disk of stone and allowed to dry at room temperature. Successive increments of H_2O were then forced through the limestone disk, and the depth of

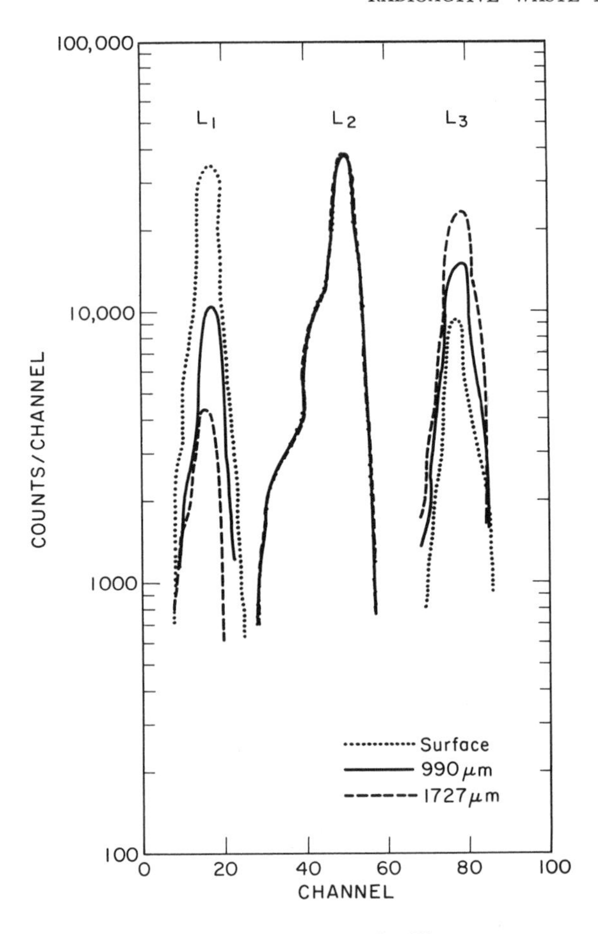

Figure 2. L x-ray spectra of the ^{238}Pu source using 9-μm-, 990-μm-, and 1727-μm-thick limestone absorbers between the source and detector. In all cases the spectra were normalized to the same number of counts in the L_2 *x-ray peak.*

penetration of the Pu tracer was measured by the x-ray absorption technique.

This technique depends on the fact that the three uranium L x-rays, which accompany 25% of all ^{238}Pu decays, are of slightly differing energies and therefore have different absorption coefficients in the limestone. Figure 2 shows the spectrum of these L x-rays observed through several samples of wet limestone. These were thin measured slices of the limestone which were interposed as absorbers between a source of ^{238}Pu radiation and the detector. The ratio of the intensities of the L_1/L_2 x-rays decreases with the thickness of the stone; similarly the L_3/L_2 ratio increases with thickness. It is a simple matter to calibrate

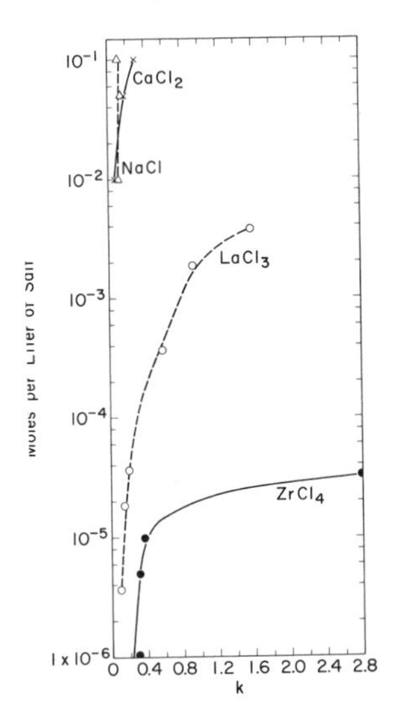

Figure 3. Surface absorption coefficients of 0.00004M $Pu(NO_3)_4$ solutions on limestone as a function of the concentration of other salts, i.e. $ZrCl_4$, $LaCl_3$, $CaCl_2$, or NaCl

the change in these ratios as a function of thickness (penetration) since the x-ray intensity ratios change logarithmically with thickness. In the time interval after each increment of H_2O had passed through the sample, the L x-ray spectrum was measured through the surface of the stone and the average depth of penetration of Pu tracer was determined by comparing the measured L_1/L_2, and L_3/L_2 ratios to the calibration curve.

The results of these experiments yielded a migration coefficient, $m = 30 \pm 10$ μm/m of water flow for the limestone and 61 ± 8 μm/m for the basalts, where m is the average distance traveled by the Pu atoms for every meter traveled by the H_2O molecules.

The second set of experiments consisted of measurements of the surface absorption coefficient of Pu on the stones. In these experiments disks of the stone were immersed in solutions of 0.00004M $^{238}Pu(NO_3)_4$. Small aliquots (0.05%) of the solutions were removed, dried on Ta planchets, and then placed in an internal alpha proportional counter. When the counting rate of samples taken at 12-hr intervals had become constant, this was regarded as evidence of equilibrium.

We will define an absorption constant as:

$$k = (\text{activity of Pu/ml of solution})/(\text{activity of Pu/cm}^2 \text{ of stone})$$

The value of k for pure solutions of $Pu(NO_3)_4$ at 0.00004M was 0.10 ± 0.02 for limestone and 0.07 ± 0.02 for basalts. To observe the

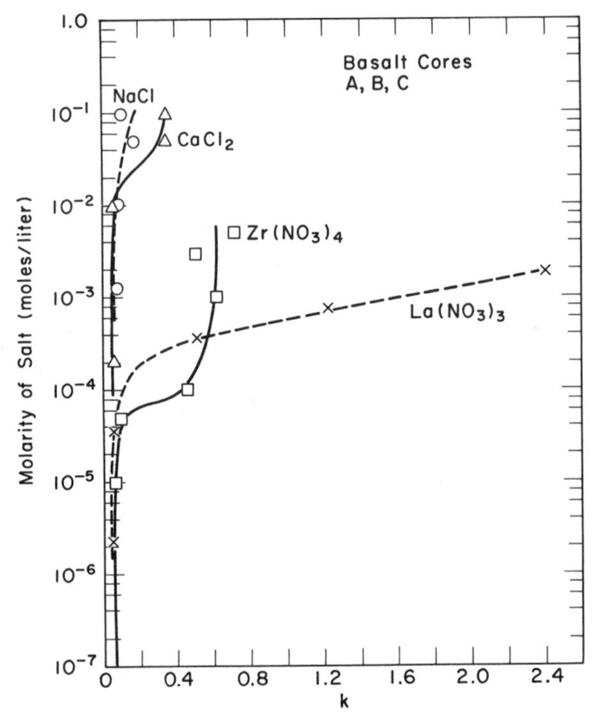

Figure 4. Behavior of the absorption coefficient of Pu as a function of the concentration of other salts for dense basalt cores

effect of other ions on this absorption constant, the value of k was measured for solutions containing NaCl, $CaCl_2$, $LaCl_3$, and $ZrCl_4$ at various molarities. Figures 3, 4, and 5 illustrate the variation in k as a function of the concentration of these salts for the limestones and basalts used. In all cases the original pH of the solution was 7.0 and the final pH ranged between 7 and 8.

Computer Model and Experimental Results with Fissure. The computer model MARGIE is a simplified variation of a calculation we have used for predicting multiorder buildup of isotopes in reactors. It traces the distribution of Pu activity over many small increments on the face of a fissure or leak during the addition and washing cycles. The only input needed for the program are the surface absorption coefficient, dimension of the fissure, and the volume of solution in which the Pu is added or with which it is eluted.

Figure 6 is an example of the output of the program for a typical calculation. Ten ml of solution containing the activity was presumed to be introduced into the top of the fissure and allowed to flow through; this was then presumed to be followed by 20 ml of water wash. The curves represent the distribution of the activity on the face of the rock

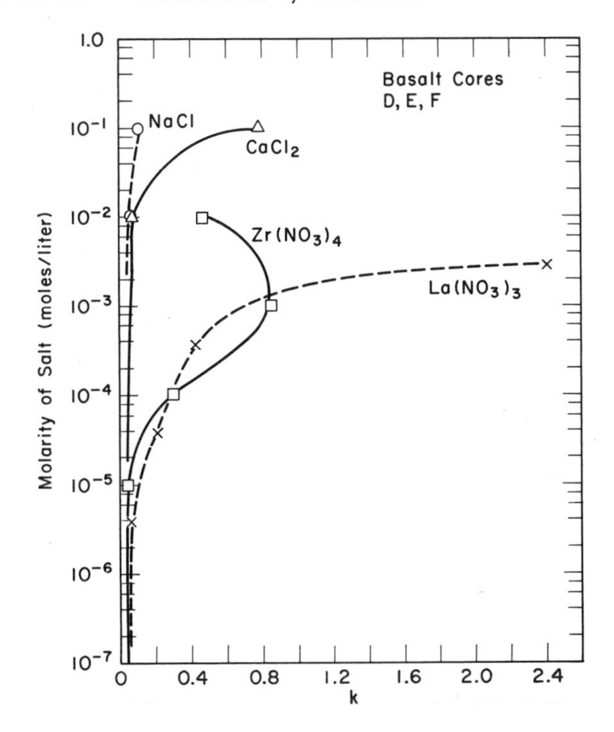

Figure 5. Behavior of the absorption coefficient of Pu for porous basalt cores

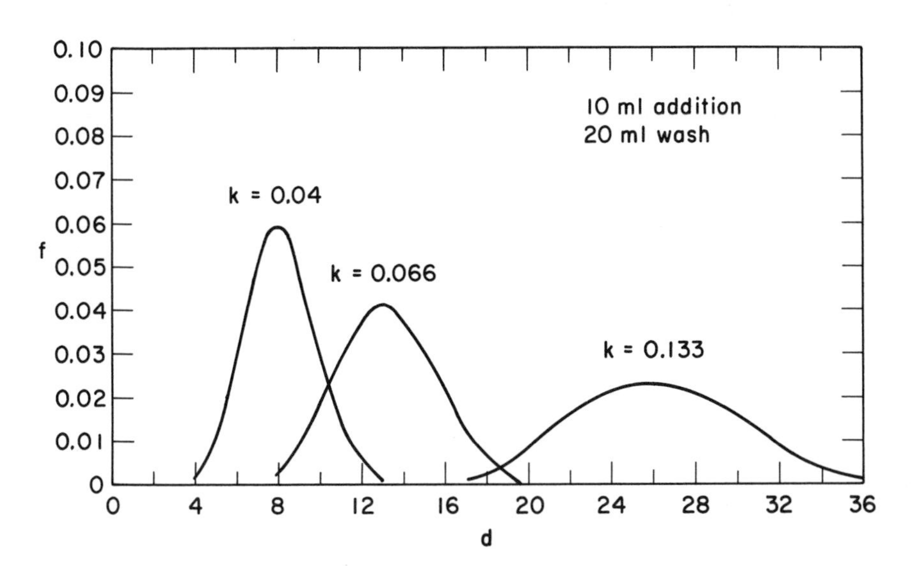

Figure 6. Calculated distribution of Pu on surface of fissure

after the wash period; it is calculated for three systems with three different absorption coefficients, $k = 0.04$, 0.066 and 0.133. In Figure 6, d is the distance along the rock face in the direction of flow expressed in millimeters, and f is the fraction of the original activity absorbed per 0.25-mm strip across the fissure. As is expected, the higher the value of k the further and more rapidly the activity moves. It should be remembered that k is defined as (activity of Pu/ml of solution)/(activity of Pu/cm^2 of rock).

To test this model of the relationship of the surface absorption coefficient to the migration rate through the fissure, we constructed a model fissure. This was done by sealing a smooth slab cut from a basalt core to the face of a Teflon block that had a 0.0127-cm depression milled down its face. This then formed a fissure that was $0.0127 \times 1 \times 4$ cm in size. ^{238}Pu tracer was dissolved in 10 ml of water and allowed to flow through this fissure at a rate of 3 ml/day. This was then followed by 20 ml of water wash. After this the slab of rock was removed and dried, the surface was scanned by a scanning alpha counter with a spatial resolution of 0.25 mm. Figure 7 is a plot of the activity observed in these 0.25-mm strips. The left hand peak at 0.20 in. (5.8 mm) from the top of the fissure corresponds to the calculated position of a peak of activity with $k = 0.04$ which is close to the value for our solutions. The

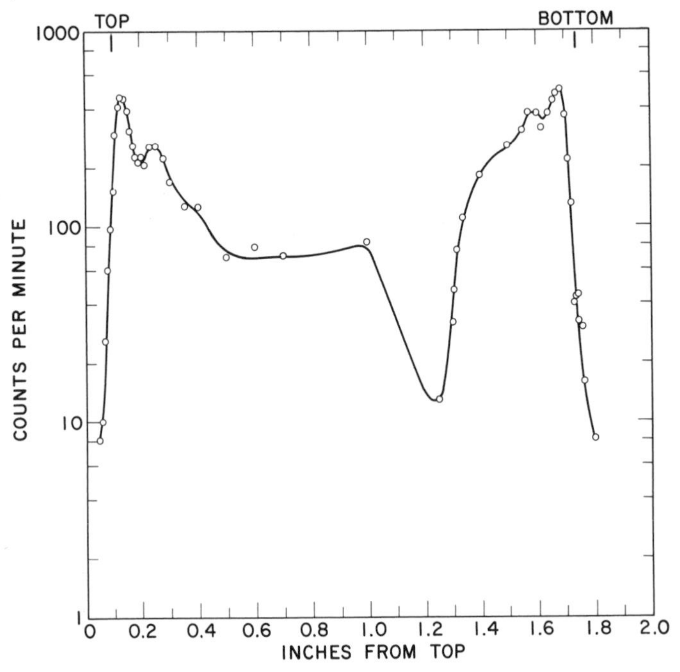

Figure 7. Measured distribution of Pu on surface of fissure

second peak at 1.7 in. (43 mm) from the top corresponds to a peak of activity with $k = 0.35$. Since the solutions were composed of pure water and tracer, this was an unexpected result.

This more rapidly moving peak, if found to be reproducible, is composed of a chemical form of plutonium; perhaps a $(Pu)_n$ polymer, which does not absorb strongly. The flow through the fissure is so fast (1.71 m/day of water flow) that it does not have time to reach equilibrium. We are now working on a method to identify the chemical form of the plutonium in this peak.

The migration of the plutonium through a fissure of this size, a 0.0127-cm crack, is much more rapid than through the small pores of the stone. Instead of 60 μm/m of water flow as found for flow through porous limestone, even the slowly moving peak ($k = 0.04$) moves 217 μm/m of water flow, and the rapidly moving peak moves more than 2000 μm/m of water flow. These rates depend on the volume/surface-area ratio of the fissure and will be larger for larger fissures. This also implies an average pore diameter of 30–60 μm for the sample of rock used in the porous flow measurements, if the migration is through interconnecting pores.

Summary

Our basic assumption of the intercorrelation of the migration coefficients and surface absorption coefficients to the migration through fissures is verified. However, a great deal of effort must be spent studying the effects of other solute species, the chemical nature of the plutonium itself, and the kinetics of the absorption process before any understanding of the macroscopic characteristics of the transport of plutonium can be reached.

RECEIVED November 27, 1974. Work supported by the U.S. Atomic Energy Commission (now ERDA).

10

Removal of Cesium and Strontium from Fuel Storage Basin Water

M. W. WILDING and D. W. RHODES

Allied Chemical Corp., Idaho Chemical Programs—Operations Office, National Reactor Testing Station, 550 Second St., Idaho Falls, Idaho 83401

Spent fuel from nuclear reactors is stored underwater at the Idaho Chemical Processing Plant for cooling and shielding before processing. The fuel storage basin water becomes contaminated with fission products, primarily cesium-137 and strontium-90, from fuel elements that "leak" and from cut pieces of fuel and miscellaneous scrap contained in cans, which are vented to release gases. This report describes laboratory research and plant-scale tests of candidate ion-exchange materials for removing cesium-137 and strontium-90 from the contaminated storage basin water, which contains moderate quantities of nonradioactive dissolved solids. Cesium-137 is removed by a zeolitic ion-exchange material; strontium-90 is removed by an organic ion-exchange resin. Operational experience with plant-size ion-exchange columns indicate that both cesium-137 and strontium-90 are removed effectively by ion exchange.

The fuel storage basin is used to store irradiated fuel elements underwater (as shielding) until they are reprocessed. This water is contaminated with cesium-137 and strontium-90 that has leaked from stored fuel.

Laboratory experiments and plant operating data indicate that cesium-137 and strontium-90 can be removed from the basin water by ion exchange even though the water contains competing cations. Zeolitic ion-exchange materials are more selective for cesium than are the organic cation resins. Zeolon-900 proved to be more selective for cesium adsorption than either of two other zeolites (clinoptilolite or AW-500) in this particular system, even though the AW-500 is reported to have a higher cation-exchange capacity than the Zeolon (*1*).

Several ion-exchange resins were candidates for the adsorption of strontium from contaminated basin water. Strontium-90 was adsorbed best by Amberlite-200 in laboratory equilibrium-type experiments and was used in plant-scale columns. Strontium was also removed by coprecipitation, using sodium oxalate to form a calcium-strontium oxalate precipitate.

Laboratory studies indicate that Zeolon-900 loaded with cesium-137 was partially regenerated with several reagents, including ammonium sulfate, ammonium nitrate, and potasium and sodium nitrates; however, regeneration of a plant-scale column was apparently unsuccessful, i.e., the concentration of cesium-137 in the effluent was approximately the same before and after regeneration.

Amberlite-200 resin loaded with strontium-90 was regenerated successfully both in the laboratory and in a plant-scale column using $5M$ $NaNO_3$. Approximately 90% of the strontium was removed using two column volumes (360 gal for 24 ft^3 of resin) of regenerant. Fifty successive regenerations have not reduced the exchange capacity of the resin.

Laboratory tests and a large-scale test in the fuel storage basin showed that strontium-90 can be removed from basin water by coprecipitating strontium with calcium oxalate. At a mole ratio of calcium plus magnesium-to-oxalate of 1.0, ca. 90% of the strontium will coprecipitate as strontium-calcium oxalate. However, the slow settling precipitate creates excessive turbidity in the water for many days, and slow hydrolysis of the strontium oxalate (ca. 10% per month) reduces the effectiveness of this treatment unless a method is available to remove the precipitate.

Laboratory tests have also shown that cesium and strontium in the effluent from the two plant-scale ion-exchange columns (Zeolon-900 and Amberlite-200) can be further removed by the use of an additional Amberlite-200 and Zeolon-900 column in series. Approximately 9000 column volumes (1.6×10^6 gal for 24 ft^3 of resin) of effluent from the first ion-exchange unit can be processed before the radioactivity concentration guide (RCG) controlled area release limit for each radioisotope is reached and before either column would need to be regenerated.

The fuel storage basin is a concrete basin ca. 20 ft deep with an operating capacity of 1.5×10^6 gal of water. It is used as an interim storage facility for a variety of irradiated fuel elements, cans containing cut pieces of fuel, and other uranium-containing scrap requiring reprocessing. The water in the basin is contaminated, primarily by fission products leaking from these stored materials.

The two major problems resulting from contamination of the water are the radiation exposure to personnel working in the area and the cost of decontaminating shipping casks before they can be transferred from the fuel storage basin facility. Prior to 1963, the contamination in the basin water was controlled by continuously purging the water to ground; the soil removed the radionuclides from the water by ion exchange. From 1963 to 1966 all the water discharged to the ground was passed through a clinoptilolite (naurally occurring zeolite) ion-exchange column, which reduced the cesium-137 and strontium-90 concentration to approximately drinking water limits (*2*).

Since 1966 the basin has been operated as a closed system. The basin water is continuously recycled through a diatomaceous earth filter, and no contaminated water is released to ground. Makeup water is added to the basin by spraying casks as a part of the cask decontamination procedure. The water level is controlled by adding additional water to the basin if needed, or by purging a small stream from the basin to the plant waste evaporator. An increase in the concentration of radionuclides and dissolved solids in the water has occurred as a result of the recycle process. Consequently, a plant-scale ion-exchange unit was installed in July 1973, based on the results of laboratory studies, to remove radionuclides from the basin water. The fuel storage basin water at that time had the approximate chemical and radiochemical composition shown in Table I.

This report describes laboratory experiments, which were used as a basis to design plant-scale equipment, and the plant-scale experience in removing cesium-137 and strontium-90 from the basin water. Cesium-137 and strontium-90 are the two major radionuclides in the water; the concentration of other radionuclides ordinarily is so low that they do not contribute significantly to personnel exposure.

A variety of cation-exchange materials was tested in the laboratory using equilibrium-type experiments and small columns to determine the capability of these materials for removing cesium-137 and strontium-90 from the fuel storage basin water.

Table I. Chemical Composition of ICPP Fuel Storage Basin Water

Chemical	*mg/l.*
Calcium	90
Chloride	320
Magnesium	17
Nitrate	200
Phosphate	< 0.04
Silicon	8
Sodium	183
Sulfate	89
Carbonate	< 1
Bicarbonate	89
Dissolved solids	905
Hardness ($CaCO_3$)	250
pH	7.4

Removal of Cesium-137

The fuel storage basin water contained ca. 0.035 μCi of cesium-137/ml when the plant-scale ion-exchange system started operation in July 1973. Cesium-137 is a long-lived (30-yr half-life) radionuclide with a 0.66 MeV gamma and is the major source of radiation exposure to personnel.

Adsorption of Cesium in Equilibrium-Type Experiments. Laboratory experiments were initiated to obtain the necessary information for ultimately designing a plant-scale ion-exchange system. Batch equilibrium experiments were used as a rapid method to select ion exchangers for testing in columns. The equilibrium experiments consisted of adding 1 g of resin to 100 ml of basin water in a stoppered centrifuge tube, then agitating the mixture at room temperature for 24 hr on a mechanical shaker. J. S. Buckingham (*3*) indicates that temperature in the 20°–40°C range has very little effect on cesium adsorption by zeolites. After 24 hr, the samples were centrifuged, and the clear supernate was analyzed for cesium-137. The distribution coefficient (K_d) was then calculated from the results as follows:

$$K_d = \frac{C_s}{C_l} \times \frac{V}{W}$$

where

C_s = concentration of metal ion in the solids

C_l = concentration of metal ion in the liquid

V = volume of liquid

W = weight of solid.

The results are tabulated in Table II.

All three zeolitic materials used in the experiments had high distribution coefficients for cesium-137. One is the naturally occurring zeolite clinoptilolite (*4*), which has the empirical formula $(M,N_2) \cdot Al_2O_3 \cdot nSiO_2 \cdot mH_2O$, where M is an alkaline earth and N an alkali metal. Another is a synthetic zeolite, Zeolon-900, with the approximate chemical composition $(Ca,Na_2,K_2)_4\ Al_8Si_{40}O_{96}\ 28H_2O$. The last is a synthetic zeolite (AW-500), which has the approximate formula $M_{12}(AlO_2)_{12}(SiO_2)_{12}$, where M is the metal ion (Na, K, Mg). Table III gives some of the physical and chemical properties of the three zeolites (*1, 5, 6*). After completion of the batch equilibrium experiments, studies were made using laboratory columns to compare the capacities of the three zeolites for removing cesium-137.

Table II. Distribution Coefficients for Cesium-137

Resin or Zeolite	K_d
Amberlite-122	787
Amberlite-200	1,350
Amberlite-525	540
ARC-359	625
AW-500	18,800
Clinoptilolite	4,750
Dowex 50-X8	2,050
Duolite C-3	2,060
Grace-544	2,520
MSC 1-H	1,000
Xe-277	200
Zeolon-900	8,000

Table III. Physical and Chemical Properties of Candidate Zeolites

Material	*Wt % H_2O at 25°C*	*Structural Type*	*Exchange Capacity (meq/g)*	*Mole Ratio SiO_2/Al_2O_3*
AW-500	15	Chabazite	2.2	4–5
Clinoptilolite	12	Clinoptilolite	1.7	8–10
Zeolon-900	12	Mordenite	1.9	10

Cesium Removal by Ion-Exchange Columns

LABORATORY STUDIES. The laboratory columns used for the ion-exchange experiments were 3/8-in. i.d. 24-in.-long tubes containing 25 ml of the resin or zeolite. The ion-exchange materials were chosen for the study primarily on the basis of the results of equilibrium studies reported in Table II. Fuel storage basin water was pumped directly from the basin and through the columns at a flow rate of 5–8 gpm/ft^2. The effluent was collected and the volume measured to determine the number of column volumes of feed passing through the column before a specified breakthrough occurred. The concentrations of cesium-137 were determined for each zeolite or resin studied. Table IV shows the breakthrough of cesium-137 at the 1% and 50% levels for the zeolites and resins tested in the column studies. Zeolon-900 was clearly the most effective of the materials tested for removing cesium-137 from basin water.

Figure 1 shows the cesium-137 breakthrough curves obtained for three zeolites in laboratory columns using fuel storage basin water as feed. The curve showing cesium-137 breakthrough for clinoptilolite was replotted from data reported earlier (*4*). Significantly, cesium-137 was preferentially removed in the presence of relatively large amounts of other cations such as Na^+, Ca^{++}, and Mg^{++}. The Zeolon-900 had more than

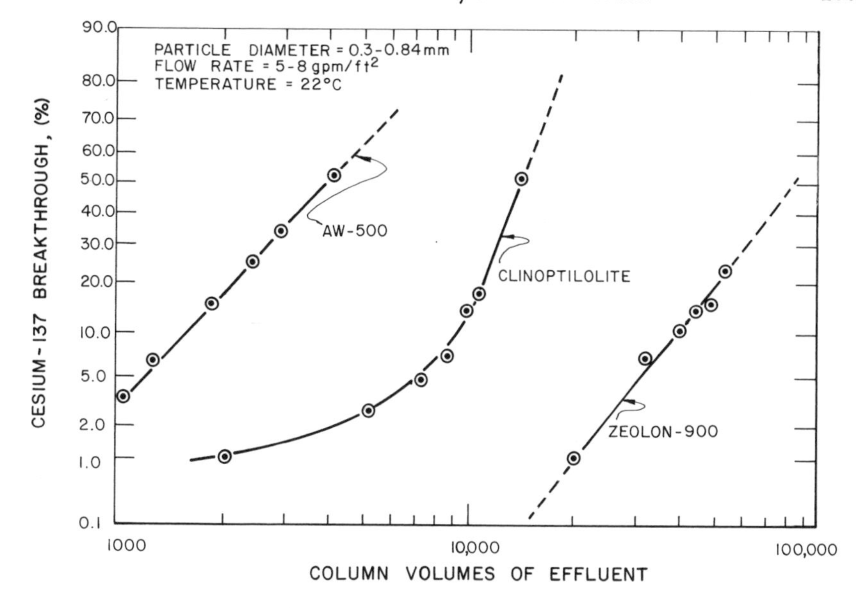

Figure 1. Cesium breakthrough curve for zeolites using basin water as feed

six times the capacity (at 50% breakthrough) for cesium-137 than did clinoptilolite, and about 21 times more than AW-500 as evident from the results shown in Table IV and Figure 1; Zeolon-900 was thus selected for use in a plant-scale column.

BREAKTHROUGH DATA FOR PLANT-SCALE ZEOLON-900 COLUMN. Based on the results of the laboratory studies, a plant-scale unit was installed to remove cesium-137 from the basin water. Zeolon-900 in the sodium form was used in the plant-scale column, which was approximately 8 ft high and 3 ft in diameter. The column was filled to a depth of 3.5 ft with 24 ft^3 of 20- to 50-mesh size Zeolon-900. Contaminated basin water was passed through the column, and samples of the column effluent were analyzed for cesium-137 to determine breakthrough values. Figure 2 shows the breakthrough curve for cesium in the plant-scale column. The 1% breakthrough volume was only 16% of that obtained in the labora-

Table IV. Cesium-137 Breakthrough for Candidate Resins and Zeolite

Resin or Zeolite Tested	*Column Volumes Resulting in Cesium-137 Breakthrough*	
	1% Breakthrough	*50% Breakthrough*
AW-500	800	4,000
Amberlite-200	250	350
Clinoptilolite	2,000	13,000
Dowex 50-X8	450	575
Zeolon-900	20,000	85,000

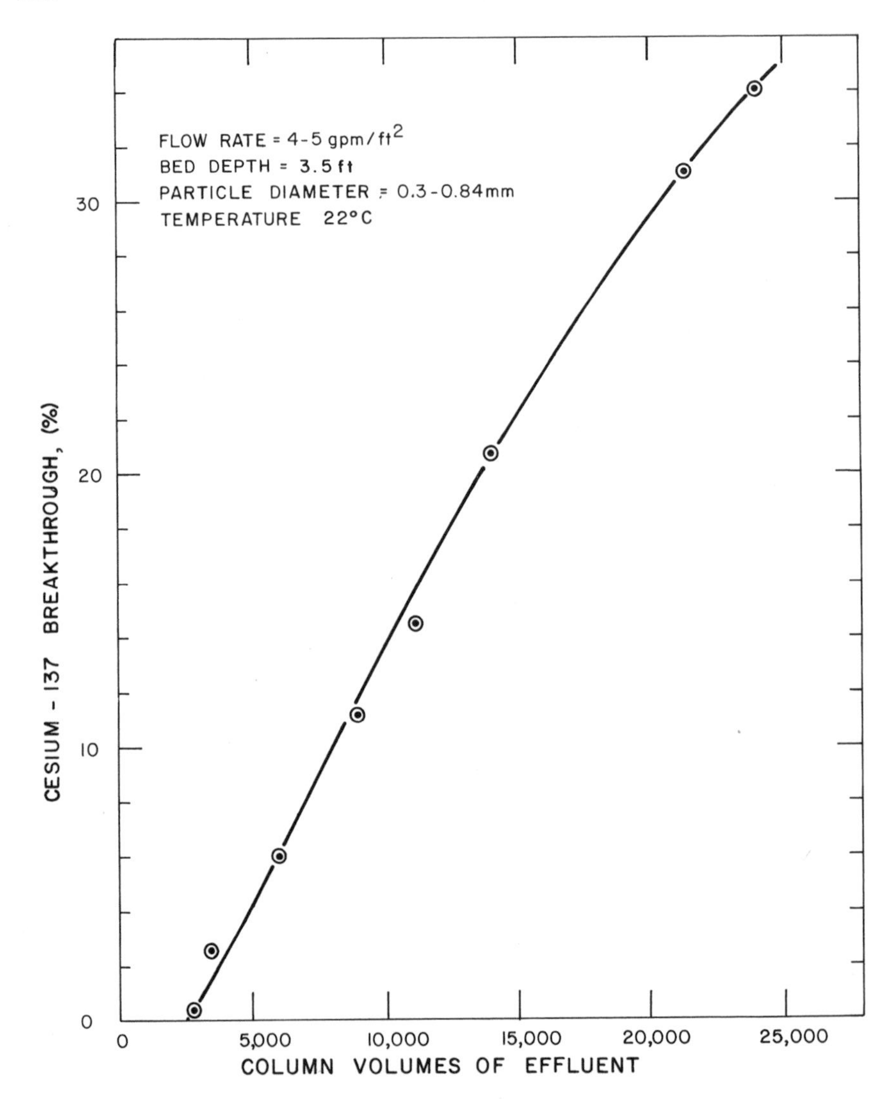

Figure 2. Cesium breakthrough curve for Zeolon-900 plant-scale column

tory column. The reasons for the lower efficiency are not known precisely; however, laboratory column operations are often difficult to apply to large-scale operations (7). Possible reasons include nonuniform flow of the liquid through the cross section of the bed, channeling, or the ion-exchange material, which may vary in efficiency from batch to batch.

Figure 3 shows the decrease in concentration of cesium-137 in the basin water as a function of time during the first six months' operation of the Zeolon-900 column. Except for a 10-day shutdown resulting from

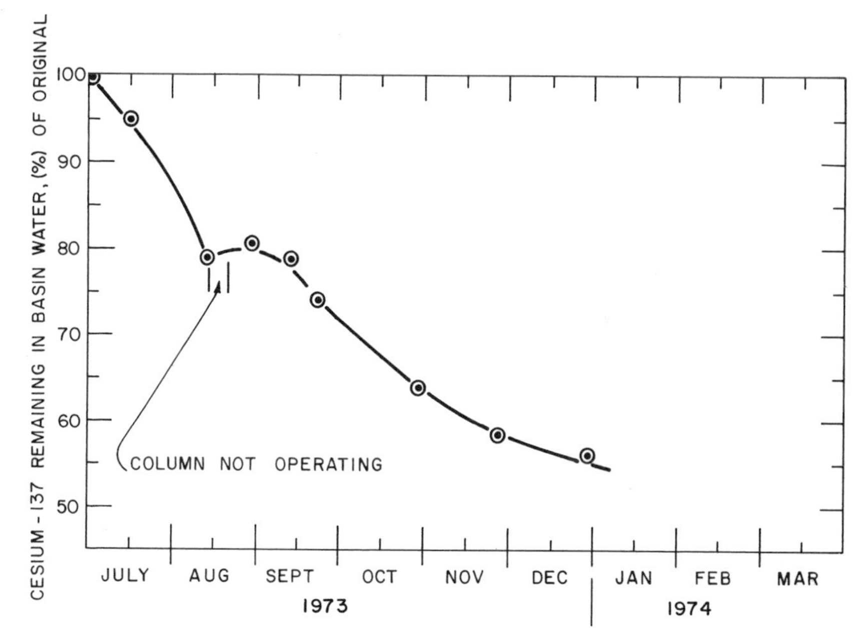

Figure 3. Cesium-137 remaining in fuel storage basin water during operation of the Zeolon-900 column

a piping change that was made in the filter system, the cesium concentration continued to decrease.

Regeneration of Zeolon-900 Column. Zeolon-900 loaded with cesium-137, which had been adsorbed from fuel basin water, was used in batch-type and column experiments to determine if the cesium could be removed by regenerating with several different reagents. The equivalent of six column volumes of regeneration solution was used for a single contact in the batch experiments. Table V shows the effectiveness of the regenerants to remove cesium-137 from the Zeolon-900 in the order of their effectiveness.

Two attempts were made to regenerate a plant-scale column using ammonium sulfate one time and sodium nitrate the second time. Although a considerable amount of cesium-137 was removed by both regenerants,

Table V. Regeneration of Zeolon-900 in Batch Experiments

Reagent	*Molarity*	*Total Cesium Removed (%)*
$(NH_4)_2SO_4$	2.5	40
NH_4NO_3	2.5	52
KNO_3	2.5	40
$NaNO_3$	2.5	24
HNO_3	2.2	16

as indicated by radiation readings of samples of the spent regenerating solution, the breakthrough value for cesium-137 (concentration in the effluent/conc. in the influent) was essentially the same before and after regeneration of the column. Because of these results, the Zeolon-900 will be replaced rather than regenerated when the concentration of cesium-137 in the effluent is greater than 50% of the concentration in the influent. Additional studies are needed to determine why the Zeolon-900 does not remove cesium effectively from basin water after regeneration.

Removal of Strontium-90

The fuel storage basin water contained ca. 0.059 μCi of strontium-90 per ml when the plant-scale ion-exchange columns started operation in July 1973. Strontium-90 is a long-lived (28-yr half-life) beta-emitter that contributes significantly to the radiation hazard for personnel working in unshielded areas where it can be concentrated, such as on the diatomaceous earth filters. Strontium-90 also makes decontamination of shipping casks difficult, primarily because of the yttrium-90 daughter, which experience has shown is difficult to wash from the casks.

Various resins and zeolites were used in equilibrium experiments and in column experiments to determine the efficiency of ion-exchange resins and zeolites in removing strontium-90 from basin water in order to select a suitable material for a plant-scale ion-exchange system.

Adsorption of Strontium in Equilibrium-Type and Column Experiments. Laboratory batch equilibrium experiments were used as a rapid method for selecting ion exchangers for testing in columns. Distribution coefficients were obtained for strontium adsorption by equilibrating 1 g of resin or zeolite in 100 ml of basin water and agitating for 24 hr at ambient temperature. After centrifuging, the concentration of strontium-90 in the supernate was determined. Table VI shows the measured dis-

Table VI. Strontium-90 Distribution Coefficients for Various Ion Exchange Materials

Resin or Zeolite Tested	K_d
Amberlite-200	340,000
Amberlite-122	8,130
Amberlite-IRC-84	150
Amberlite-IRC-50	100
ARC-359	13,000
AW-500	2,700
Clinoptilolite	490
Dowex 50-X8	15,000
Duolite C-3	2,480
IRA-938	22
Nalco HDR-W	7,240
Zeolon-900	18,000

Table VII. Strontium-90 Breakthrough Values Using Laboratory Ion Exchange Columns

Resin or Zeolite	*Column Volumes*	
	1% Breakthrough	*50% Breakthrough*
Amberlite-200	1,200	1,800
AW-500	40	150
Clinoptilolite	150	2,000
Dowex 50-X8	800	1,000
Zeolon-900	10	100

tribution coefficients. Based on these results, five ion exchangers were selected for testing in laboratory columns to obtain capacity values. In these experiments, fuel storage basin water was passed through ⅜-in.-dia. by 24-in.-long tubes containing 25 ml of resin or zeolite at a flow rate of 5–8 gal/min/ft^2 and at ambient temperature. Aliquots of the effluent were collected and the concentration of strontium-90 was determined in these samples. The results of the column experiments are shown in Table VII. Based on these results, Amberlite-200 resin was selected for use in the plant-scale column.

Removal of Strontium-90 from Basin Water in a Plant-Scale Column. Amberlite-200 was selected for use in the plant-scale ion-exchange column because (a) a large volume was processed (1200 column volumes) before 1% breakthrough occurred, (b) a reasonably large volume was obtained before 50% breakthrough (a measure of capacity) occurred, (c) it is a macroreticular resin that is resistant to physical degradation, and thus provides high flow rates, and (d) it is readily available at a reasonable cost.

A large plant-scale column (24 ft^3 of resin) of Amberlite-200 was installed to remove strontium-90 from the effluent of the Zeolon-900 column. Figure 4 shows a typical breakthrough curve obtained for strontium-90 using the plant-scale column. The apparent capacity (30% breakthrough) of this plant-scale column was about 900 column volumes, which was about one-half of the value obtained in the laboratory columns. Figure 5 shows the decrease in strontium-90 concentration in the fuel storage basin water over a six-month period while the column was operating. Approximately 250,000 gal of fuel storage basin water are processed before 100% breakthrough of strontium-90 occurs, at which time the column is regenerated.

Regeneration of Amberlite-200 Resin. The Amberlite-200 resin was operated in the sodium form during the exchange cycle; strontium and calcium ions replace sodium ions. In the regeneration cycle, high concentrations of sodium ion must be used to displace the strontium and calcium. Sodium nitrate was used to regenerate laboratory columns

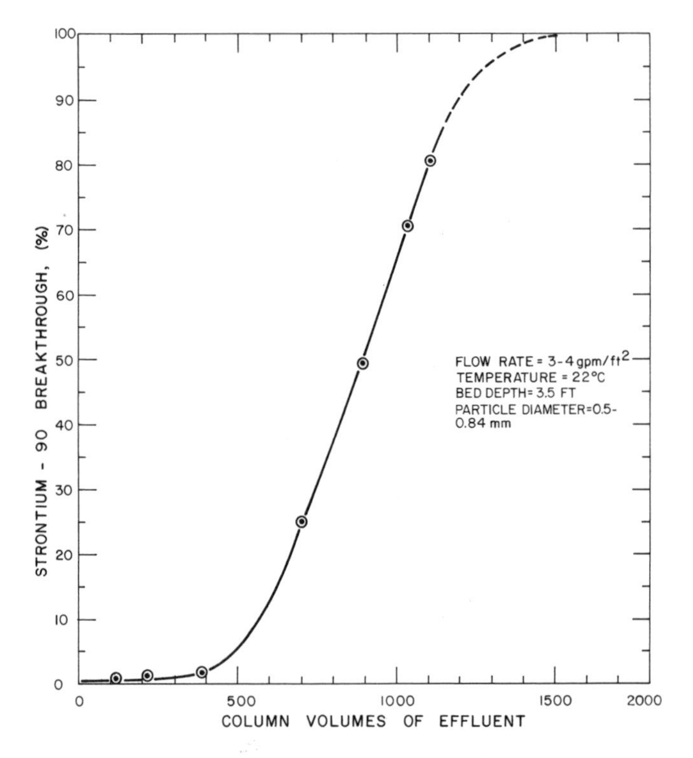

Figure 4. Strontium breakthrough curve using Amberlite-200 resin in a plant column

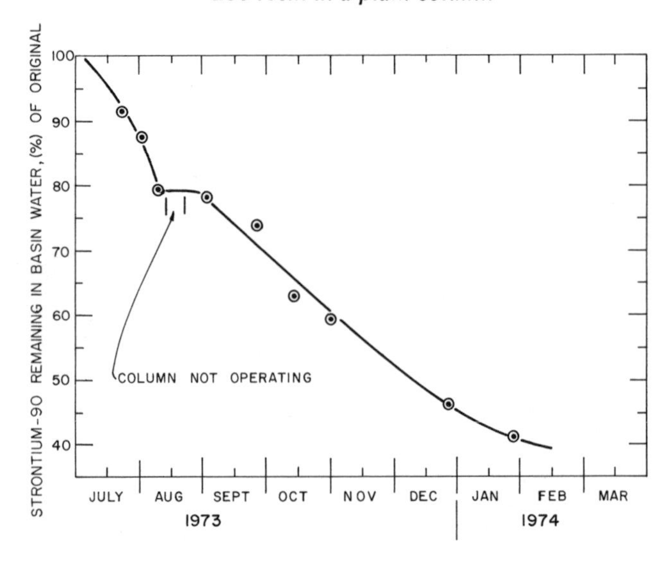

Figure 5. Strontium remaining in fuel storage basin water during operation of the Amberlite-200 column

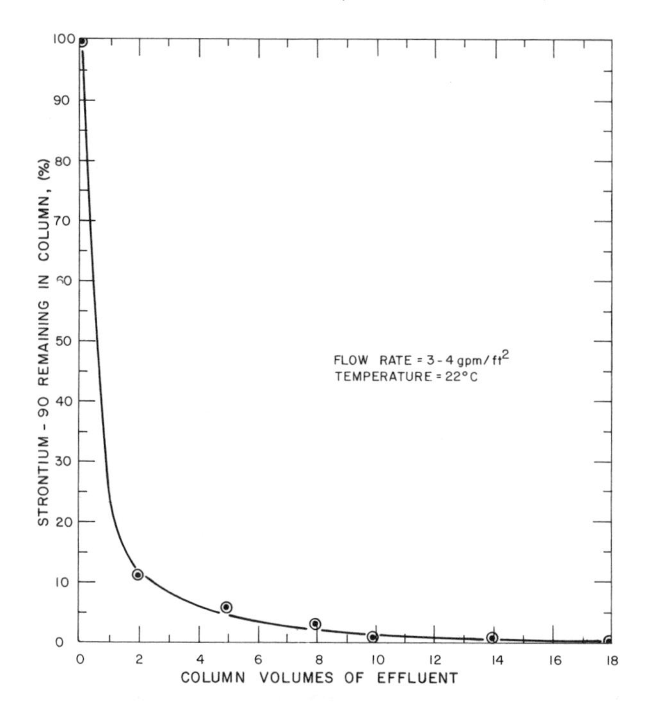

Figure 6. Regeneration of Amberlite-200 with $NaNO_3$ in a laboratory column

loaded with strontium-90 and calcium from fuel storage basin water. A solution of 5*M* $NaNO_3$ was passed through the column containing the Amberlite-200 at a flow rate of 3–4 gpm/ft^2. Tests were made at both ambient temperature and at 70°C. The results are shown in Figure 6. Approximately two column volumes of regenerant were required to remove 90% of the strontium-90 from the resin.

The plant-scale ion-exchange column containing Amberlite-200 has been regenerated 50 times. The regeneration has been approximately 90% effective for each regeneration cycle, using approximately 1.5 column volumes of sodium nitrate per regeneration. The regeneration efficiency in the plant-scale column is about the same as that observed in laboratory columns. The resin appears to have approximately the same exchange capacity after 50 cycles as it did initially, as indicated by the low concentration of strontium-90 in the column effluent when the adsorption cycle is started after regeneration.

Strontium Removal by Precipitation

Calcium and magnesium are present in the fuel storage basin water at a concentration of approximately 90 and 17 ppm, respectively. Cal-

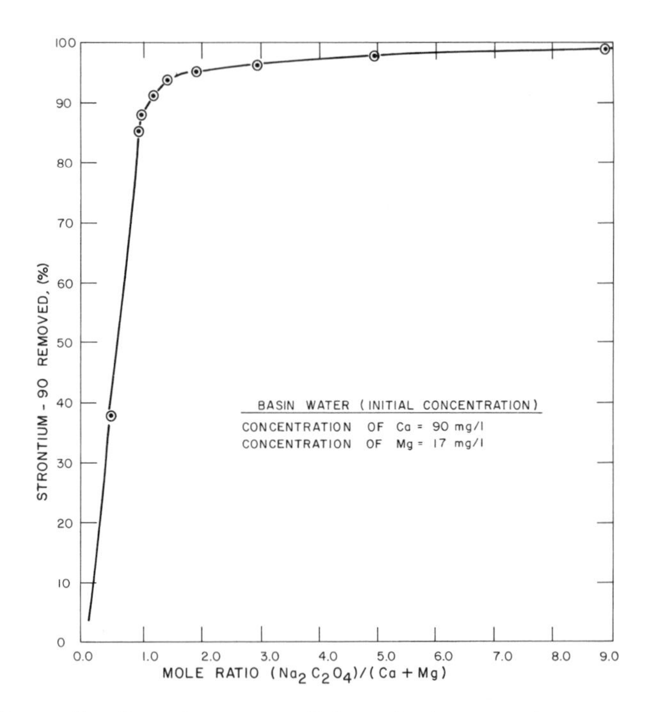

Figure 7. Strontium removal from basin water by coprecipitation with calcium oxalate

cium and magnesium compete for the exchange sites in the ion-exchange resin and therefore limit the capacity of the resin to adsorb strontium. Precipitation of calcium, and to a lesser extent magnesium, from the water would therefore increase the onstream time for the ion-exchange column and at the same time reduce the concentration of strontium-90 in the water if the strontium-90 coprecipitates with the calcium. Therefore, a series of laboratory experiments was made to determine the effectiveness of a precipitation method for reducing the concentration of strontium-90 and calcium in basin water.

Laboratory Tests. Attempts were made to precipitate calcium and strontium from fuel storage basin water, using both oxalic acid and sodium oxalate. Although both reagents were effective, sodium oxalate was selected because the crystals dissolved rapidly and the pH of sodium oxalate solution (pH 7.4) is compatible with the basin water.

The mole ratio of sodium oxalate-to-calcium plus magnesium was varied from 0.5 to 9.0 in order to determine the minimum amount of oxalate needed to precipitate the strontium. The results are shown in Figure 7. Approximately 90% of the strontium was removed at a mole ratio of 1.0; above a mole ratio of 2.0, little additional strontium was removed.

In addition to the laboratory tests, 50 gal of basin water were treated in a 55-gal stainless steel drum to simulate treatment of the basin with sodium oxalate. A mole ratio of oxalate-to-calcium plus magnesium of 1.25 was used. The results were equivalent to those obtained using 100-ml laboratory samples; more than 90% of the strontium and calcium was removed. Based on the favorable results of these tests, the water in the fuel storage basin was treated by adding granular sodium oxalate to the water.

Treatment of Fuel Storage Basin Water with Sodium Oxalate. Sodium oxalate was added to the water as a granular solid and was distributed by hand over the same area several times to ensure an even distribution of the chemical over the area treated. Sufficient sodium oxalate was added to provide an oxalate-to-calcium plus magnesium ratio of about 0.5. Approximately three weeks were required for the precipitate to settle; apparently thermal currents created by heat emanating from the stored fuel elements slowed the settling process considerably, because preliminary tests indicated that the precipitate would settle to the basin floor in about three days. Samples of the fuel storage basin water were taken periodically to determine the amount of strontium and calcium removed by the treatment. The results indicated that about one-half of the calcium and strontium precipitated, which was in good agreement with results obtained in the laboratory. A second identical treatment was planned but was not carried out because of the long settling time required before the water was sufficiently clear to handle fuel. The need to transfer fuel made it impractical to take the basin out of service long enough for a second treatment.

Although the solubility of the precipitated calcium oxalate is low—0.0067 g/l. at 13°C (8)—laboratory experiments were run to determine if dissolution occurred which would slowly release strontium back into

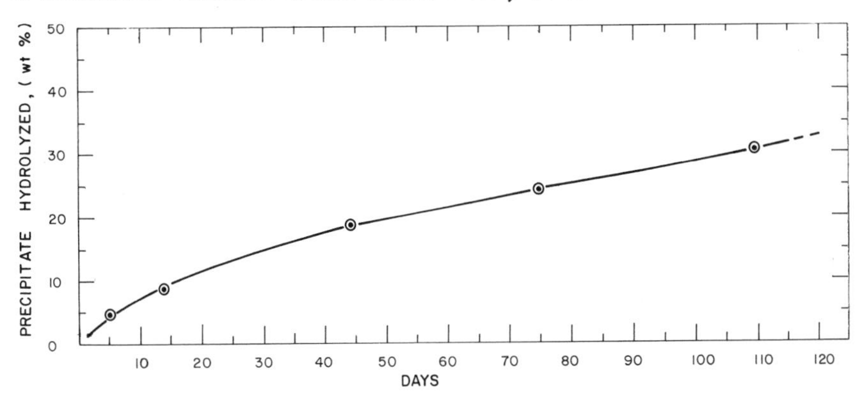

Figure 8. Dissolution of calcium-strontium oxalate precipitate in fuel storage basin water

solution. The laboratory results, which are shown in Figure 8, show the rate of dissolution of the calcium-strontium oxalate in fuel storage basin water. The apparent rate of dissolution is about 10% per month indicating that the precipitate should be either removed by filtration or vacuumed from the floor of the basin as soon as possible. A vacuum system has been installed at the fuel storage basin to remove sludge from the floor; however, inoperability of the sludge pump has prevented removal of the sludge to date.

Decontamination of Fuel Storage Basin Water to Controlled Area Discharge Limits

The concentration of dissolved salts in the fuel storage basin water has been increasing steadily since recycle was started in 1966. Chloride buildup is the primary concern; sources of the chloride are: water used to wash the casks, and the chlorine treatment used to inhibit the growth of microorganisms. Chloride is potentially corrosive to aluminum fuel elements. One method considered for reducing the concentration of dissolved salts in the fuel storage basin water would be to reduce the concentration of cesium-137 and strontium-90 in the water to below the RCG release limits (9) so that the water can be released to ground. Up to 0.1 Ci of cesium-137 and strontium-90 may be releaesd to the ground

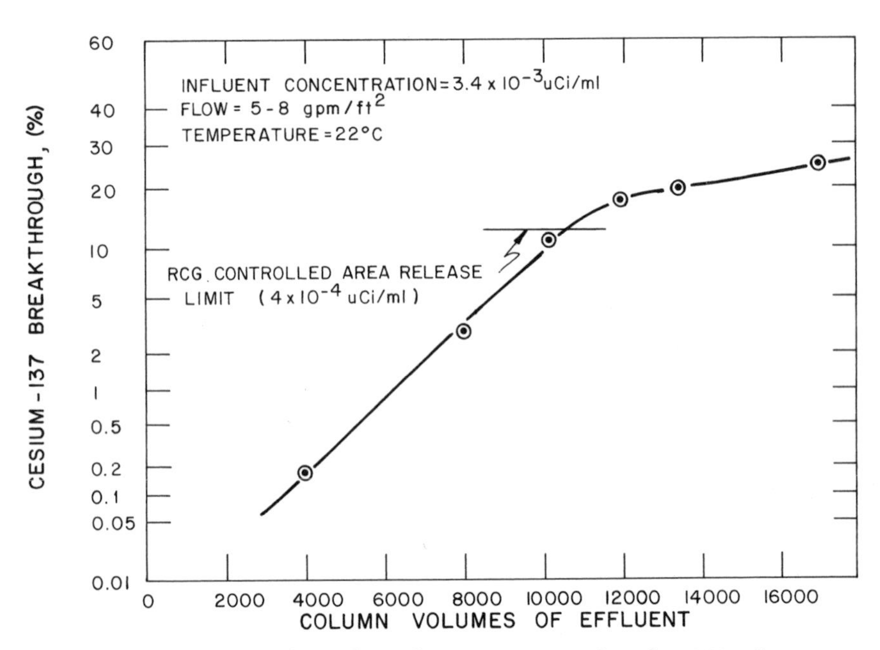

Figure 9. Cesium breakthrough curve for second Zeolon-900 column

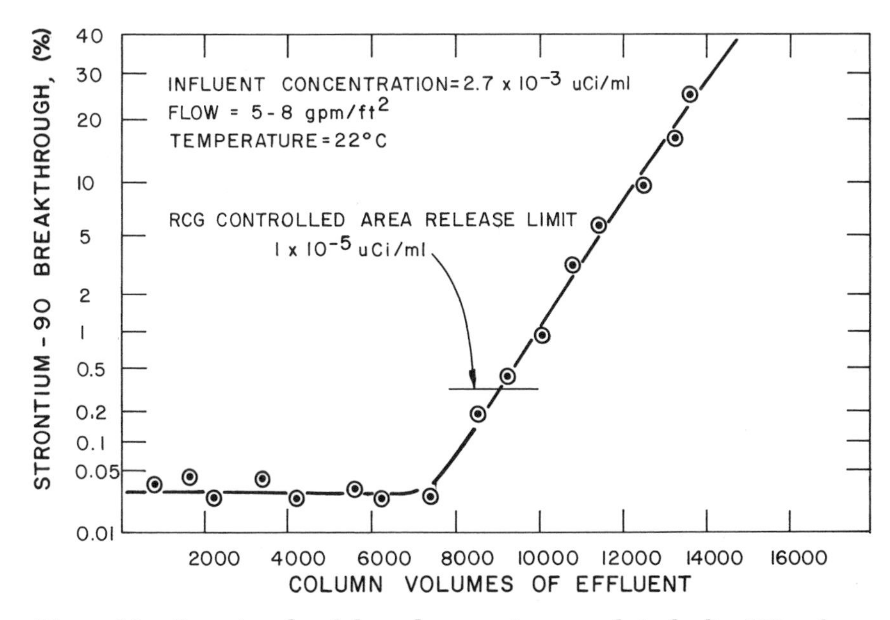

Figure 10. Strontium breakthrough curve for second Amberlite-200 column

on a "one-time" basis if this technique were employed. The cesium-137 and strontium-90 concentration in the effluent from the existing ion-exchange unit is higher than allowable release limits. Approximately 3.2×10^6 gal would need to be processed to reduce the concentration of chloride from 400 to ca. 50 ppm, which is considered a tolerable concentration.

A laboratory experiment was run using the effluent from the existing plant-scale ion-exchange columns as feed to a Zeolon-900 column and to an Amberlite-200 column. This effluent feed contained about 3.0×10^{-3} μCi/ml of strontium-90 and about the same concentration of cesium-137. Figure 9 shows the breakthrough curve for cesium-137 in the second Zeolon-900 column. Approximately 11,900 column volumes were processed before the RCG controlled area release limit for cesium-137 (4×10^{-4} μCi/ml) was reached. Figure 10 shows the breakthrough curve for strontium-90 in the second Amberlite-200 column. Approximately 9000 column volumes were processed before the RCG controlled area release limit of 1×10^{-5} μCi/ml for strontium-90 was reached. These data indicate that if two additional ion-exchange columns (one containing Zeolon-900 and the other Amberlite-200) were operated in series with the existing plant-scale ion-exchange columns, ca. 9000 column volumes (1.6×10^6 gal) of effluent from the existing ion-exchange system could be processed and released to ground without exceeding the RCG controlled area release limits, and before the second set of columns would need regeneration. This method of reducing the concentration of dissolved salts in a second ion-exchange system is dependent on the effectiveness of the first ion-exchange system, which in turn is dependent on the rate of fission product leakage from the stored fuel. Another approach being considered for removing chloride from water, which is not dependent on the concentration of radionuclides, is demineralization by ion exchange. This method would use a cation exchange resin in the hydrogen form, which would be regenerated, and an anion resin in the hydroxide form, which would be buried as a solid waste when the ion-exchange capacity is exhausted.

Conclusions

Based on the results of laboratory experiments and plant operating experience, the following conclusions were reached. (a) Cesium-137 is selectively removed from water containing the competing calcium, magnesium, and sodium cations by the zeolite, Zeolon-900. Thus, a high capacity (ca. 30,000–40,000 column volumes is estimated for a plant-scale column) is attainable. Regeneration of the Zeolon-900 in a plant-scale column was not successful. (b) Strontium-90 is removed from water by Amberlite-200 resin, but calcium and magnesium are also removed, which results in a low capacity for strontium-90 (900 column volumes for a plant-scale column). Regeneration with sodium nitrate was successful. Regeneration of the Amberlite-200 resin 50 times did not affect the ability of the resin to remove strontium-90. Strontium-90 may also be removed by coprecipitation of strontium oxalate with calcium oxalate, but the precipitate should be removed by filtration if good water clarity is required. (c) Additional treatment in a second ion-exchange system in series with the first unit can reduce the concentration of cesium-137 and strontium-90 to controlled area discharge limits, thus permitting release of the water to the ground to dispose of excessive dissolved solids.

Based on the preceding conclusions, the following actions are underway. (a) Instead of regenerating the Zeolon-900, it will be replaced periodically—the time for each cycle to depend on the "removal goals." (b) The Amberlite-200 column will continue to be used with frequent regeneration until a significant decrease in capacity occurs. (c) More studies will be made to determine why the plant column containing Zeolon-900 did not regenerate. This should include a study of the exchange mechanism and the factors affecting the removal of cesium-137 on Zeolon-900 and AW-500.

Literature Cited

1. Ames, L. L., Knoll, K. C., "Loading and Elution Characteristics of Some Natural and Eynthetic Zeolite," **HW-74609**, 1962.

2. Rhodes, D. W., Wilding, M. W., "Decontamination of Radioactive Effluent with Clinoptilolite," **IDO-14657,** July 1965.
3. Buckingham, J. S., "Laboratory Evaluation of Zeolite Material for Removing Radioactive Cesium from Alkaline Waste Solutions," **ARH-SA-49,** January 1970.
4. Wilding, M. W., Rhodes, D. W., "Removal of Radioisotopes from Solution by Earth Materials from Eastern Idaho," **IDO-14624,** 1963.
5. Mercer, B. W., Jr., Ames, L. L., Jr., "The Adsorption of Cesium, Strontium, and Cerium on Zeolite from Multication Systems," **HW-78461,** August 1963.
6. Mercer, B. W., Jr., Ames, L. L., Jr., "Adsorption of Cesium and Strontium on Zeolites from Multicomponent Systems," **HW-SA-3199,** November 1, 1963.
7. Dorfner, K., "Ion Exchangers, Properties and Applications," Ann Arbor Science Publ., Inc., 1972, pp. 121–122.
8. Weast, R. C., Ed., "Handbook of Chemistry and Physics," 54th ed., Cleveland: CRC Press, 1973–74.
9. U.S. Atomic Energy Commission, "Standards for Radiation Protection," from "AEC Manual," Nov. 8, 1968, Chapter 0524.

RECEIVED November 27, 1974. Work done under contract AT(10-1)-1375 S-72-1.

11

Determination of Transuranics in Environmental Water

NORBERT W. GOLCHERT and JACOB SEDLET

Argonne National Laboratory, Argonne, Ill. 60439

One of the requirements of any nuclear facility is to monitor the effluent waste water to show compliance with existing standards. This paper describes a sequential procedure for the separation of the transuranic elements from water samples up to 60 l. The elements of interest are coprecipitated with calcium fluoride and then individually separated using a combination of ion exchange and solvent extraction, with a final sample preparation by electrodeposition. Alpha spectrometry of these samples allows the measurement of neptunium, plutonium, and transplutonium nuclides at sub-fCi/l. levels.

Facilities that process high-level radioactive material are required to demonstrate that their radioactive emissions are below standards set by regulatory agencies, or do not otherwise present a health hazard. This is usually accomplished through an environmental monitoring program to show what, if any, nuclides leave the site. In contrast to the situation inside these facilities, where very high concentrations of radioactive materials exist, the concentrations in the environment are very low and frequently require large samples for detection.

This paper discusses that specific area of a monitoring program that deals with the determination of transuranium nuclides in water. The initial work in this area was to measure the concentration of plutonium in large volume water samples (*1*). This method has been extended to include the determination of other transuranium elements.

The procedure involves the coprecipitation of the transuranic nuclides on calcium fluoride from acid solution after reduction of the plutonium and neptunium with bisulfite. The calcium fluoride precipitate is dissolved in aluminum nitrate–nitric acid solution and the plutonium and neptunium separated on an ion-exchange resin column. The column effluent solution is extracted to separate the transplutonium nuclides. The separated fractions are electrodeposited and the concentrations determined by alpha spectrometry. The chemical recoveries are determined by adding known amounts of plutonium-236, neptunium-239, and americium-243 for the transplutonium nuclides.

Experimental

Reagents. The anion-exchange resin was AG-1-X8, nitrate form, 100–200 mesh, BIO-RAD Laboratories, Richmond, Calif. The aluminum nitrate–nitric acid solution was prepared by saturating 8*N* HNO_3 with reagent grade $Al(NO_3)_3 \cdot 9H_2O$. This solution was purified by passing it through a nitrate form AG-1 resin column. The extractant, Aliquat-336 (methyl tricaprylyl ammonium chloride), was obtained from General Mills, Inc., Kankakee, Ill. as the chloride salt. A 30% (v/v) solution was prepared in xylene and converted to the nitrate salt by mixing equal volumes of 30% Aliquat-336 and 1*M* $Al(NO_3)_3$ for 5 min. This was repeated three times with fresh portions of $Al(NO_3)_3$ solution.

Sample Pretreatment. Water samples are acidified with nitric acid immediately after collection to inhibit hydrolytic loss of trace elements and then filtered to remove suspended solids. Surface water samples frequently contain suspended soil from runoff water. This soil contains some transuranium nuclides, particularly plutonium, from fallout of previous atmospheric nuclear tests. The amount of suspended soil in water may be substantial under certain conditions. This pretreatment is used to obtain the concentration of the soluble transuranium elements. The filtered residue can be analyzed if desired.

Procedure. The procedure is outlined in Figure 1. The first step is the separation of the desired nuclides by coprecipitation with calcium fluoride. The optimum nitric acid concentration for effective carrying on calcium fluoride is between 0.1*N* and 0.2*N*. Reduction of the plutonium to Pu(III) was necessary to obtain quantitative carrying on calcium fluoride. Plutonium(IV) is known to form colloidal or "non-ionic" species in neutral solution, and in this form may be incompletely carried by calcium fluoride. Bisulfite was effective and gave complete reduction in 3.5 hr at 50°C or overnight at room temperature. The concentration of calcium must be at least 0.1 mg/ml for quantitative carrying.

Calcium fluoride will coprecipitate most actinides and lanthanides, but few other elements. The lanthanide elements will separate with the transplutonium fraction. The concentration of the lanthanides in environmental water is at the sub-microgram per liter level (*2*). The mass of lanthanide elements electrodeposited from a typical sample does not interfere with alpha spectrometric energy resolution. Although rare earth

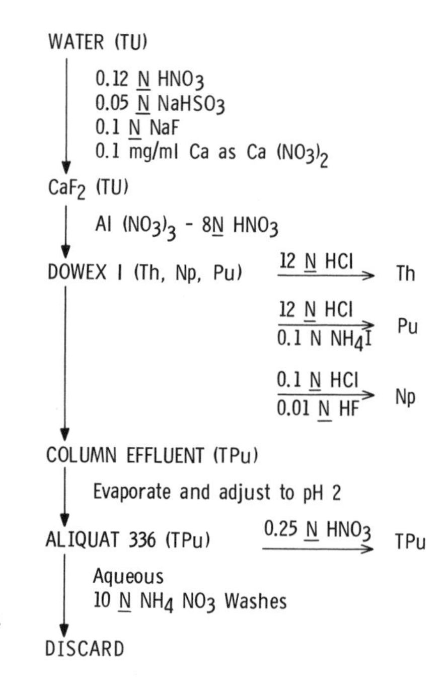

Figure 1. Transuranium separation procedure

fluorides are equally effective as calcium fluoride in carrying the actinides, the alpha radioactivity of rare earth salts can be substantial and varies not only with the rare earth but also between lots of the same compound.

Dissolution of the calcium fluoride in aluminum nitrate–nitric acid oxidizes the plutonium to the tetravalent hexanitrate complex (*3*), while the transplutonium nuclides remain in the trivalent state. The only actinides retained by a nitrate-form anion-exchange column are thorium, neptunium, and plutonium. The uranium distribution coefficient under these conditions is about ten, but uranium should not be present at this point since hexavalent uranium does not carry on calcium fluoride (*4*).

The thorium is removed from the column with 12*N* HCl, since it does not form a chloride complex, to separate it from neptunium and plutonium. The plutonium is removed by reduction to Pu(III) with 12*N* HCl–0.1*M* NH_4I solution. The neptunium is then removed by dilute acid. The neptunium and plutonium can also be eluted together with dilute acid and the various nuclides determined by alpha spectrometry.

The column effluent solution containing the transplutonium nuclides is evaporated, adjusted to pH $\sim$ 2, and extracted with 30% Aliquat-336. The organic phase is scrubbed with 10*M* NH_4NO_3 to remove the residual calcium and aluminum, and the transplutonium nuclides are stripped into dilute acid. Water samples up to 60 l. have been analyzed by this procedure.

Discussion

The sensitivity for measurement of these nuclides is a function of a number of parameters. For our systems, typical parameters are: spectrometer counting efficiencies of 30–35%, counting times of 3000–4000 min, spectrometer background plus reagent blanks of 3–7 counts per thousand min depending upon which nuclide is considered; chemical recoveries of about 85% for neptunium, greater than 90% for plutonium, and 80% for the transplutonium nuclides; and a sample volume of 10 l. Using these parameters, and a definition of the detection limit as a net sample counting rate equal to the background plus reagent blank, the values in Table I were calculated. To check the detection limits experimentally, plutonium-239 at a concentration of 0.6 fCi/l. was added to demineralized water and was readily detected by this method. The measured concentrations were within 15% of the concentration added.

The differences in sensitivity are due to differences in the counter background plus reagent blank. It is important to minimize the contribution to the blank of radioactive impurities in the reagents. Unpurified aluminum nitrate will double the reagent blank for plutonium-239. The purification of the aluminum nitrate is described in the reagents section. Although the transuranic nuclides will coprecipitate on preformed calcium fluoride, use of the commercial reagent increases the plutonium-239 blank by a factor of ten. Neptunium-237 has essentially the same alpha energy as uranium-234 and thorium-230, which are present in small amounts in many reagents and counter materials, thereby resulting in a higher counter background in this energy region. Higher counter backgrounds also exist for plutonium-238, but this is caused primarily by the presence of a small amount of plutonium-238 in the plutonium-236 used

Table I. Detection Limits

Nuclide	*Alpha Energy (MeV)*	*Principal Interference*	*Alpha Energy of Interference (MeV)*	*Detection Limit (fCi/l.)*
Neptunium-237	4.787	Thorium-230	4.684	2
		Uranium-234	4.774	
Plutonium-238	5.499	Americium-241	5.486	2
		Thorium-228	5.424	
Plutonium-239, 240	5.155	—	—	0.5
Americium-241	5.486	Plutonium-238	5.499	1
		Thorium-228	5.424	
Curium-244 and/or	5.806 and/or	Plutonium-236	5.767	1
Californium-249	5.810	Radium-224	5.684	
Curium-242 and/or	6.112 and/or	Bismuth-212	6.089	1
Californium-252	6.118			

to determine plutonium chemical recovery. Actually, plutonium-242 is a superior isotope to use for determining plutonium chemical recovery, but its alpha energy is similar to that of neptunium-237 and would interfere with this determination. If the measurement of neptunium-237 is not desired, plutonium-242 is the isotope of choice.

Several studies were carried out to examine the effect of the colloidal or "non-ionic" form of plutonium in the analysis of water samples by the method described in this paper. The colloidal or polymeric plutonium was prepared as described by Lindenbaum (*5*). Polymeric plutonium was added to pH 7 water and allowed to stand 3 weeks. Analysis by this method recovered all of the added plutonium and indicates that this procedure is effective in determining plutonium even if present in the polymeric form.

Polymeric plutonium was added to pH 7 water and allowed to stand about one month in a glass bottle. The water was poured into a clean bottle and analyzed as described. Only 25% of the plutonium was found in this water. The original bottle was refilled with water and this method used to analyze the water while in the original bottle. The remaining 75% of the plutonium was found in the second water solution. This indicates that polymeric plutonium in neutral solution could be easily lost on container walls, but was readily removed by this procedure.

This procedure has been used in the analysis of environmental water samples over the past 3.5 yr. Plutonium measurements have been made since mid-1971, neptunium-237 since mid-1972, and transplutonium since mid-1973. The application of this procedure to routine surveillance can be illustrated by previously published results for several transuranium nuclides (*6–9*).

An attempt was made to determine a baseline concentration of these nuclides in environmental water. Our results to date indicate that all the transuranium nuclides, except plutonium-239, are below the detection limit of this procedure. Analysis of smaller bodies of water, particularly rivers and streams, shows a seasonal variation in the plutonium concentration. This is the "spring maximum" due to the stratospheric fallout. In 1973 concentrations ranged from the detection limit of 0.5 fCi/l. to 1.0 fCi/l. and averaged 0.7 fCi/l. Samples of 45 l. were analyzed for some locations, including Lake Michigan where concentrations ranged from 0.5 to 0.9 fCi/l. Corresponding concentrations have been found in the Lake by Wahlgren and Nelson (*10*). A typical sample of Pacific Ocean surface water, collected in June 1971, was reported to have a plutonium-239 concentration of 0.72 fCi/l. (*11*).

This procedure is designed to provide analytical data to show compliance with existing standards. In addition, it can be used to determine baseline concentrations of the transuranium nuclides in the environment.

These data will be valuable for comparison purposes if the proposed expansion of the nuclear industry proceeds as planned and the use of some of these nuclides, e.g., plutonium-238 and californium-252, becomes widespread.

Acknowledgment

This work was performed under the auspices of the U.S. Atomic Energy Commission (now ERDA).

Literature Cited

1. Golchert, N. W., Sedlet, J., *Radiochem. Radioanal. Lett.* (1972) **12,** 215.
2. Weast, R. C., Ed., "Handbook of Chemistry and Physics," The Chemical Rubber Co., Cleveland, Ohio, 48th ed., 1967, F-136.
3. Ryan, J. L., *J. Phys. Chem.* (1960) **64,** 1375.
4. Grindler, J. E., "The Radiochemistry of Uranium," **NAS-NS-3050,** 1962, 58.
5. Lindenbaum, A., Westfall, W., *Int. J. Appl. Radiation Isotopes* (1965) **16,** 545.
6. Sedlet, J., Golchert, N. W., "Environmental Radioactivity at Argonne National Laboratory, Annual Report for 1971," July, 1972. (Prepared for the U.S.A.E.C. and available from the authors.)
7. Sedlet, J., Golchert, N. W., Duffy, T. L., "Environmental Monitoring at Argonne National Laboratory, Annual Report for 1972," U.S.A.E.C. Report **ANL-8007,** Mar, 1973.
8. Sedlet, J., Golchert, N. W., Duffy, T. L., "Environmental Monitoring at Argonne National Laboratory, Annual Report for 1973," U.S.A.E.C. Report **ANL-8078,** Mar, 1974.
9. Sedlet, J., Golchert, N. W., Duffy, T. L., "Environmental Monitoring at Argonne National Laboratory, Annual Report for 1974," AEC/ERDA Report **ANL-75-18,** April, 1975.
10. Wahlgren, M. W., Nelson, D. M., "Plutonium in Lake Michigan Water, Argonne National Laboratory Radiological and Environmental Research Division Annual Report, Jan–Dec, 1972," U.S.A.E.C. Report **ANL-7960,** Part III, 1973, p 7.
11. Wong, K. M., Hodge, V. F., Folsom, T. R., "Proceedings of Environmental Plutonium Symposium," E. B. Fowler, R. W. Henderson, and M. F. Milligan, Eds., U.S.A.E.C. Report **LA-4756,** 93, 1971.

RECEIVED November 27, 1974.

12

Radioactive Waste Management Development in Europe

RAY D. WALTON, JR.

U. S. Energy Research and Development Administration, Wash., D.C. 20545

Germany, England, and France are vigorously developing technology and methodology for incorporating high-level radioactive waste into silicate glass. Germany is concentrating on a spray calcination vitrification system. England has selected a rising level glass process in which evaporation, calcination, and borosilicate glass vitrification all take place in a heated pot. France is operating a small-scale batch pot calcination–batch vitrification system and developing a new continuous system using a rotary calciner and melter with batchwise draw-off.

During the period May 4–18, 1973, several European sites were visited to observe waste management facilities and to discuss waste management development programs. Of primary interest was the technology for vitrification of high-level radioactive waste.

The United Kingdom

The UK has spent several years developing the FINGAL Process and comparing it with other processes for vitrification of high-level waste. In this semicontinuous process liquid waste and borosilicate glass-making constituents are slowly added to a heated pot. Three layers exist in the pot—a top liquid layer, a middle calcine layer, and a bottom molten glass layer. This is termed a rising level glass process because the glass level continues to rise during the process until the pot is filled to a designated level.

Process and equipment development will continue at Harwell with a concentrated effort on a modification of the FINGAL Process called HARVEST which features an annular calcining and vitrification unit.

This unit will be 13 ft high, with an external diameter of 30 in. and an internal diameter of 16 in. Resistance heating will be used for this annular space unit, and the design is primarily aimed at increasing the amount of heat input through the walls of the vessel. All gases will be treated to prevent the deposition of ruthenium prior to entering a condenser/absorber where it, as well as a major fraction of the oxides of nitrogen, will be removed by a caustic scrubber; the gases will then be dehumidified and filtered prior to discharge to the atmosphere.

The Federal Republic of Germany

In the FRG a great deal of attention is being given to basic chemical and physical research and development associated with glass-making processes. A process based on spray calcination at 450°C followed by semicontinuous incorporation in glass with batch drawoffs of glass-waste product, to be demonstrated in a hot cell facility, is the primary candidate. Nonradioactive process testing is nearing completion, and radioactive testing is expected to be initiated early in 1974. In the calciner, heat is supplied by recycled superheated steam to eliminate hot walls and to reduce greatly the volume of off-gas. Filters are used to separate solids from the off-gas stream, with periodic blowback of superheated (400°C) steam to keep the filters clean. The current water balance around the spray calciner-melter is one l. of high-level waste in and two to three l. of low-level contaminated water out. This is no problem in the current system as a low-level evaporator is available. However, for a production unit it appears that this balance will have to be brought into line.

Considerable effort is being expended in determining the characteristics of the waste glasses. Very sophisticated instrumentation is being used for differential thermal analysis and gravimetric thermal analysis of the waste glasses. One of the basic German thoughts is that devitrification or crystallization will take place during storage at high temperatures and in a radiation field. This crystallization may increase the solubility of the glass. However, by special heat treatment a microcrystalline ceramic material similar to Pyroceram can be formed. The Germans feel that this ceramic is superior to glass for very long storage of material. Thus, specially controlled cooling and heating cycles are being investigated in order to produce the superior microcrystalline structures.

A thermite process with continuous addition of calcined waste and other dry constituents to the reactor vessel has also been developed on a nonradioactive basis. In a demonstration run approximately 2 kg of dry ingredients reacted in about 2 min at a temperature of approximately 2000°C in a ceramic crucible.

France

France has developed and is operating on a small scale a process which involves batch calcination followed by batch mixing with borosilicate glass additives, and melting to incorporate the waste in a borosilicate glass. All Marcoule high-level waste, mostly from gas-cooled reactor fuel irradiated to approximately 4000 Mwd/MTU and highly concentrated because it is relatively free of impurities such as iron or nitrate salts, is being processed through a pilot plant. The waste glass is being drawn into SS pots approximately 35 cm in diameter and 0.5 m high which are covered but not sealed. These containers are transported in a shielded container to the nearby storage facility which is designed for 500 containers and is only 18 m square. Twenty containers of glass are stacked on top of each other in a vertical underground shaft 45 cm in diameter and 13 m long with a floor plug at the surface. Forced air, 25,000 m^3/hr with velocities up to 4 m/sec, is used for cooling. Although there are redundant systems to ensure that there will not be a loss of cooling air flow, the worst possible consequence was assumed to be total melting of the glass.

Current development is directed toward a continuous rotary calciner directly feeding into a borosilicate glass melter with batchwise drawoff of the waste glass.

Full-scale nonradioactive engineering demonstration units are being tested. In order to reduce the requirements for radioactive tests, these units are designed for remote operation and maintenance.

Summary

In summary, Germany, England, and France are vigorously developing technology and methodology for incorporating high-level radioactive waste in silicate glass.

RECEIVED November 27, 1974.

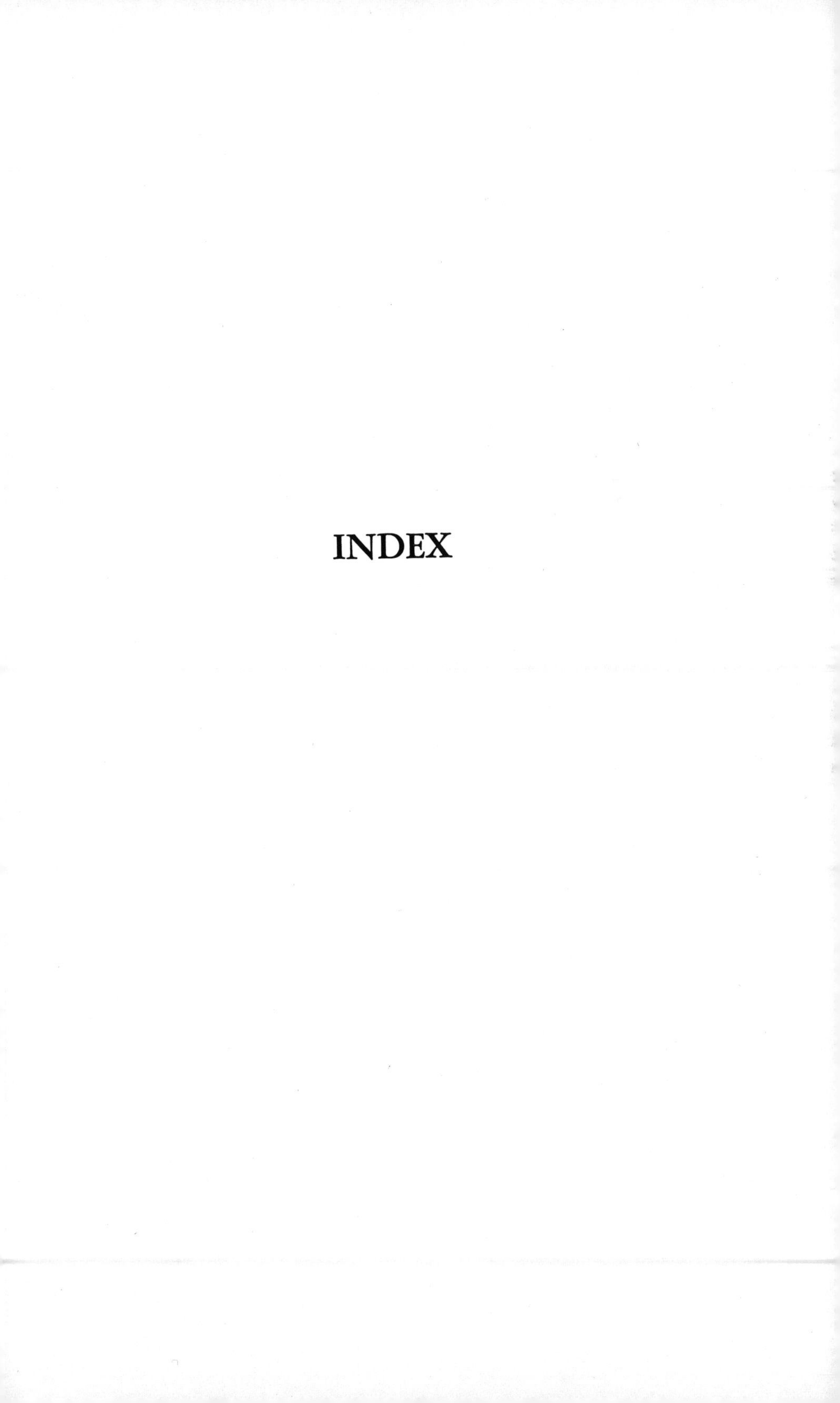

INDEX

INDEX

F

G

H

I

K

L

M

N

O

P

The text of this book is set in 10 point Caledonia with two points of leading. The chapter numerals are set in 30 point Garamond; the chapter titles are set in 18 point Garamond Bold.

The book is printed offset on White Decision Opaque 60-pound. The cover is Joanna Book Binding blue linen.

Jacket design by Linda Mattingly. Editing and production by Joan Comstock.

The book was composed by the Service Composition Co., Baltimore, Md., printed and bound by The Maple Press Co., York, Pa.